5/0

The Tribe of Tiger

The Tribe of Tiger
Cats and their Culture

Elizabeth Marshall Thomas

ILLUSTRATED BY JARED TAYLOR WILLIAMS

WEIDENFELD & NICOLSON
LONDON

First published in Great Britain in 1994
by Weidenfeld & Nicolson

The Orion Publishing Group
Orion House
5 Upper Saint Martin's Lane
London WC2H 9EA

First published in the United States of America
in 1994 by Simon & Schuster

ISBN 0 297 81508 3

A catalogue record for this book is available from
the British Library

Printed and bound in Great Britain by
Butler & Tanner Ltd, Frome and London

For Stephanie, for Ramsay

About the Title

THE TITLE OF THIS BOOK IS FROM A POEM CALLED "REJOICE in the Lamb," written sometime between 1756 and 1763 by the English poet Christopher Smart. The poem is long and rambling to the point of incoherence, a product of the confusion the poet experienced and for which he was kept in solitary confinement in a madhouse. A more frightful setting than a rat-infested madhouse of the eighteenth century would be hard to imagine, as would the loneliness and despair that Smart must have known during his eight-year ordeal. His torment was mitigated, however, by the presence of a cat, Jeoffry, who became the subject of one section of the poem—some seventy-five radiant lines that are today well known and much beloved by cat fanciers and that often appear as a poem in their own right, usually under such titles as "Of His Cat, Jeoffry" or "Jeoffry" or "For Jeoffry, His Cat." The rest of the poem is virtually lost, known only to a handful of scholars of English literature.

My book is but one of dozens, perhaps even hundreds, of books about cats that take their titles from "Jeoffry." Almost anyone who reads the fragment, even those who are unaware of Smart's confinement and suffering, can share the strength of his feeling. In the following lines, for example, one feels the

7

poet's prayerful gratitude for Jeoffry's company in the echoing asylum during the black, terrifying hours of night:

> For I will consider my cat, Jeoffry.
> For he is the servant of the Living God, duly and daily serving him.
> For he keeps the Lord's watch in the night against the adversary.
> For he counteracts the powers of darkness by his electrical skin and glaring eyes.
> For he counteracts the Devil, who is death, by brisking about the life.

One feels how the cat touched the poet's heart:

> For having considered God and himself, he will consider his neighbor.
> For if he meets another cat he will kiss her in kindness.
> For when he takes his prey he plays with it to give it a chance.
> For one mouse in seven escapes by his dallying.

And one feels the poet's inspiration:

> For he is of the Tribe of Tiger
> For the Cherub Cat is a term of the Angel Tiger.

I chose to take a title from the poem mainly because it represents a powerful link between a person and an animal, and also because the poem expresses the sanctity of an animal who the poet feels is serving God by his wholesome behavior and is keeping the devil at bay. I also chose the poem for its excellent and touching insights, made centuries before anyone thought that animals were deserving of good observation. To call the light brushing of noses between two cats a kiss, for instance, is to describe perfectly the greeting of cats who know each other (suggesting, incidentally, that there may have been a cat population as well as a rodent population in the asylum) while the observation that Jeoffry lost one mouse in seven because he played with his prey is worthy of a modern field biologist. But most of all, I chose the poem because it expresses something that is intensely true of and important about cats—that their tribe is the tribe of tigers. As the cherub is to the angel, so the cat is to the tiger, and although today we tend to put the relationship the other way around, saying that tigers are a kind of cat rather than that cats are a kind of tiger, the fact is that cats and tigers do represent the two extremes of one family, the alpha and omega of their kind.

Introduction

ONE SUMMER EVENING AT OUR HOME IN NEW HAMPSHIRE, MY husband and I were startled to see two deer bolt from the woods into our field. No sooner had they cleared the thickets than they stopped, turned around, and, with their white tails high in warning, looked back at something close to the ground as if whatever frightened them also puzzled them. We were wondering aloud what might be threatening the deer when to our astonishment our own cat sprang from the bushes in full charge, ears up, tail high, arms reaching, claws out. The deer fled, and the cat, who fell to earth disappointed, watched them out of sight.

Our cat is a male and at the time was just two years old. He weighed seven pounds and stood eight inches at the shoulder, in contrast to his two intended victims, who weighed more than a hundred pounds apiece and stood three feet at the shoulder. Even so, the difference in size and the difficulty of the task seemed to mean nothing to our cat. We realized we hadn't known him.

In truth, most of us don't know our cats. Hunting, if we stop to think of it, should of necessity be topmost in their minds. Not that the obsession always shows for what it is—sometimes its manifestation is obscure. As an adolescent, this particular cat, for instance, became enormously excited by a winter storm. Eyes blazing, he tore around in circles through the whirling snow, leaping high to catch the flying leaves that the wind was ripping from the oak trees. In addition, he forcefully tackles just about everything that comes his way, including a large loaf of Italian bread my husband had removed from a shopping bag and placed on the kitchen counter. To our amazement, the cat came hurtling up from the floor, landed on top of the loaf, bestrode its back with his claws dug deep into its sides, and instantly delivered what should have been a killing bite to the "nape" of its neck on the left side. When the loaf took no notice, the cat swarmed all over it, slashing at it with his claws and biting it deeply all over its body. Still the loaf ignored him. Suddenly the cat stopped short, stared down at his prone quarry, and then, perhaps reasoning that the loaf was dead after all, quickly scratched around it as if covering it with leaves, as wild cats hide the uneaten portions of large victims from competitors.

But there were neither leaves nor competitors in the kitchen. Abandoning the idea, the cat turned back to the loaf and rubbed it twice with his lips, swiping first to the left and then to the right. Then he jumped down from the counter and without a trace of confusion or embarrassment strolled from the room, head and tail high. Later, I photographed the badly mauled loaf of bread with the killing bite still showing on the nape—the very place, according to field studies, where most man-eating tigers, such as those of the Sundarbans in the Ganges Delta, seize their victims.

The cat has a tiger's name, Rajah (in just about every collection of zoo or circus tigers, one is named Rajah), and formerly had been called Rajah Tory Peterson because of his interest in birds. But after he proved himself willing to hunt everything from deer to loaves of bread, we dropped the surnames. Later it became clear that Rajah was not the only cat in our community to hunt so ardently—not five miles from our home live other cats who also hunt deer, and have done so long enough that the resident deer have lost all fear of them. At the sight of little cats staring up at them from the long grass the deer whistle and stamp their feet in angry warning, but to no avail. The moment they lower their heads to eat, the determined cats creep nearer.

Like Rajah, these cats are well cared for by their owners, whose homes they share and whose affection they return. Not one of these cats must catch his own food in order to eat. What then explains the obsessive hunting? The answer to that question is the subject of this book—a story of cats and the cat family.

PART ONE

The Animal

The story of cats is a story of meat, and begins with the end of the dinosaurs. Before their mysterious disappearance, the dinosaurs had reached a sort of climax in the art of meat-eating, which had begun simply enough, almost with life itself, when the early swarms of small aquatic creatures had little else to eat except one another. For these early swimmers, plants as we know them were not an option, since plants had not evolved. As life became more complicated, hunting and meat-eating became more complicated too. Most of the vertebrates were meat-eaters—certainly most of the fish ate other fish, as did the first amphibians, who in turn became food for the emerging reptiles.

During Permian and Triassic times, predatory dinosaurs crowded out most of their meat-eating forebears, ending the long reign of the big carnivorous amphibians. From Jurassic times onward even the largest dinosaurs had predatory dinosaurs trying to kill them, with more dinosaurs waiting to scavenge the remains. The mammals had no chance to mount any kind of challenge. As a result, when after 130 million years of highly successful predation the dinosaurs vanished, they left behind a most unusual situation—a world newly free of carnivores of any appreciable size.

Even as recently as the Paleocene, sixty-five million years ago, only two groups of mammals could have been called carnivorous, and by today's standards, or indeed by the former high standards of the predatory dinosaurs, neither of these would have seemed particularly adept at hunting. The first group, called the creodonts, were not built for speed and probably specialized in carrion, and the second group, called the miacids, ancestors of the modern carnivores, were for the most part very small and possibly specialized in insects; in other words, despite the new situation, both groups continued to eat what they had been eating when the dinosaurs were still around.

Thus, there was no one to molest the millions of large, hairy, milk-fed animals who soon evolved to roam the fertile forests left them by the dinosaurs, browsing the trees and bushes without much fear of predators. The early herbivores became a vast, slow-moving food supply which eventually even they themselves could not ignore. A few of them, including an enormous hoglike, bull-sized creature called *Andrewsarchus*, gave up leaves for meat. The largest carnivorous mammal ever to have lived on land, this minotaur with a carnivorous habit probably lacked the delicate sensibilities of a true carnivore such as a dog or cat or weasel and surely must have been the most frightening predator the world has ever seen.

By all that is sensible, the creodonts should also have been developing themselves to better exploit the world of meat. Surprisingly, however, they went into decline. Their eventual extinction seems puzzling—not only had some of them grown to the size of bears, so that they were much heftier than their insect-eating competition, but they were in the very act of evolving larger brains and learning to run faster. Even so, they disappeared.

And thus, a way opened for the modern carnivores. Encouraged by the magnificent opportunities and in the absence of rivals, the former little insect-eaters grew and changed. From them emerged two bloodlines, sometimes known as the Vulpavines, or Fox Tribe, and the Viverravines, or Mongoose Tribe. (Students of Latin should simply ignore the fact that *viverra* means ferret, since a ferret is not a kind of mongoose at all but a kind of weasel and belongs with the dogs in the Fox Tribe. Also best ignored is that *viverra* comes from *wer*, the Indo-European root word for squirrel. Squirrels aren't carnivores, of course, and except as prey have no place in this story.) During the Oligocene, members of the Fox Tribe started turning into the dogs, the bears, the raccoons, and the weasels. Members of the Mongoose Tribe became the modern mongooses, the hyenas, and the cats.

While these new carnivores became the hunters of the plains and forests, and also, by Miocene times, hunters of the sea as walruses, sea lions, and seals, many of the Fox Tribe secured for themselves a place close to the evolutionary middle ground. Although they were carnivores, they may have slowly weaned themselves from their insectivorous habits by eating some vegetable matter too. Insects and plants, after all, are so intermixed that they can almost be called two parts of the same thing, and go together like franks and beans, so to take a bite of one is often to nip a little of the other by mistake, especially if whoever is eating has a big mouth. If, like most animals, you have no hands to pick up your food but must put your mouth directly on it, when you snap up an insect, you sometimes can't help biting off a piece of the plant it was sitting on. Conversely, if you eat the plant, you often chew up some insects accidentally. (Allegedly it's hard to tell the difference. A man who ate a caterpillar in a desperate, failed effort to impress a woman said later that the caterpillar had tasted like the plant he'd found it on.)

To the early Fox Tribe, the ability to eat vegetables proved very helpful in times of meat shortages. Today some of its members—the raccoons and bears, for instance—can eat almost anything. This allows them to exploit a wide range of habitats. Others, such as the dogs and modern foxes, can endure a decline in their usual meat supply by varying their diet to include fruits, legumes, insects, and other forms of nourishment. One very early member of the bear family, the bamboo-eating giant panda, has passed right through meat-eating into vegetarianism again and thus has become, paradoxically, a vegetarian carnivore.

The cats, however, took a riskier path. In general, the descendants of the Mongoose Tribe eat fewer vegetables than do the descendants of the Fox Tribe, and the cats eat almost none. A cat might eat the chyme that had been in a victim's stomach, or take a little catnip as a recreational drug, or chew some leaves for vitamins, or swallow a few sharp blades of grass as a scour, but cats can't extract enough nourishment from these or other vegetables. Well-meaning human vegetarians notwithstanding, cats must eat animal protein or they slowly decline and eventually starve. Not for them the comfortable middle ground, eating meat one day and berries the next, and no carrion either. Fresh meat killed by themselves or by their mothers is virtually the only item on the feline menu. The cats have chosen the edge.

Survival at the edge is no easy matter. The food of cats is not found at the tips of branches, waiting aromatically in the sun to disappear down someone's throat. Fruit, after all, is the reward offered by a plant to anybody willing to swallow its seeds, soften the husks, and eventually put them on the ground, securely packed in fertilizer and too far away from the parent plant to offer any competition. In contrast, the food of cats is frightened of the cat and is dedicated to its own survival. It is intelligent, brave, fast moving, often well armed, and sometimes much larger than the cat who wants it. So to live at the edge, the cats were challenged to become highly skilled as hunters.

And this they did. Hunting preoccupies a cat almost from birth. The behavior of kittens at play is hunting behavior and nothing else. Because a cat can hunt without eating but cannot eat without hunting, hunting means life to cats, so much so that the process of hunting matters more than the resulting food. A cat of ours, named Orion because of his unquenchable hunting (and also because he seemed to say "Orion" when he gave his ringing, far-traveling call), brought no fewer than thirty chipmunks into our house during just one summer. Head high, pace determined, jaws bulging with the chipmunk who was forced to ride, feet forward, in his mouth, Orion would hurry to the living room, put down his victim, and step back.

Of course, once the poor creature got its bearings it would try to escape, and Orion would chase it from room to room. Up and down the curtains the chipmunk would run with the cat leaping after him, up and down the stairs, over the beds, under the sofa, over the kitchen table, through the sink and out into the hall. The moment my husband and I would hear what the cat was doing we would of course join the chase in hopes of rescuing not only the chipmunk but also our things. Even though we almost always managed to catch the chipmunk and let it out, thus depriving Orion of his sport and his prey, he nevertheless seemed to enjoy our participation, possibly because we added to the excitement, or possibly because housecats (contrary to what many people believe) do in fact hunt cooperatively when the opportunity arises. They don't cooperate as well as lions do, but they cooperate to some extent—a practice that seems to have developed from the tendency of kittens to follow their hunting mothers, trying to take part as best they can. At any rate, perhaps in hopes of our participation, Orion released chipmunks again and again, to the point where we hated to hear the cat door slam because we knew what was coming.

Interestingly enough, our area could not possibly have sustained the thirty chipmunks captured by Orion. At most, ten or twelve might have lived in our woods and stone walls. This meant that Orion was catching some chipmunks for a second or third time. One chipmunk lost a tail, and two were slightly wounded by Orion's inconsiderate games, but only three of his victims died. And not from any lack of hunting ability on the part of the great Orion, who enjoyed his sport for one summer and then gave it up for reasons of his own but not from a lack of chipmunks.

Had he been cruel? Well, yes, by human standards. But human standards mean little to the cats. Furthermore, incredible as it seems, chipmunks are hunters too—not as skilled as the cats, of course, but equally cruel. Once one of the above-mentioned chipmunks caught a wood frog and ate it, swallowing one of its legs and packing the rest of it, still feebly struggling, into his cheek pouches. The frog was almost half the size of the chipmunk, so it didn't fit in easily. Three or four times the chipmunk was forced to drag the dying frog out of his mouth, turn it around, and repack it. So Orion, who may have captured this same chipmunk once or more than once, seemed almost kind by comparison.

Throughout the cat tribe, many individual cats can kill without benefit of experience or education, contrary to an often-stated belief that killing is a skill that mother cats (but not father cats) teach their young. Mother cats certainly are teachers, but exactly what they must accomplish with this teaching is imperfectly understood, at least by human beings. A female puma, for instance, who was born in a zoo and knew less than most of us about the ways of the wild, instantly killed an unfortunate young male elk whom her keepers had found injured by the side of a road and, in order to make a video of their puma killing something, had shoved into her pen.

Even more revealing, perhaps, is the account of a puma named Ruby, who was born on a fur farm but rescued as a tiny kitten by her owner, a wildlife rehabilitator, Lissa Gilmour. Except for her first few weeks of life, Ruby had always lived with Lissa at Lissa's home in Colorado, and was completely uneducated from a puma's point of view. Thus it was impressive to learn how much information on hunting was already in Ruby's head.

One evening, Lissa was giving a lecture on the puma at the Denver Museum of Natural History, with Ruby scheduled to make an entrance after the slides had been shown. Friends had offered to help by keeping Ruby occupied until the time came for her to join Lissa, at which point they were to bring her to the lecture hall. The kindly people were doing exactly that, and Ruby, restrained by her collar and a leash, was patiently padding along beside them through the dark, deserted halls of the museum, when in one of the dioramas she spotted a stuffed deer. Instantly she sprang at it, whisking the leash out of her handler's grasp. Alas, she crashed into the glass and dropped to the floor, so the experience, for her, must have been quite bewildering. Yet for the rest of us it must be considered extremely illuminating, since Ruby had never done any such thing before. She had never seen any animal killed, let alone a deer. Furthermore, the deer in the exhibit certainly hadn't moved to attract her, nor had it given off a tempting sound or odor. No—Ruby had reacted to its appearance only, and her reaction had been sure and strong.

These episodes show something important about the cat family—that meat-eating is deeply ingrained in their nature. Consistent meat-eating explains much about all cats, from why, except for size and camouflage, there is very little difference among the thirty-two species of the family to why they seldom mark with feces but frequently use urine, which they spray.

In short, cats resemble each other because, so far, they have had no reason to change. Good hunters since the lynxlike Urcat of the Miocene from whom the modern cats descend, the cats have had no need to adjust their bodies or their diets in response to major changes in the world's climate. Why not? Because, unlike the diets of other animals, the diet of cats didn't change. The vegetarian menu listed everything from bananas to pecans, from seaweed to eucalyptus leaves, items so different from one another that completely different organisms were required to find, chew, and digest them, but the cat menu listed only one item: meat. From a cat's point of view, the difference between a bird who eats cherries, a fish who eats algae, and a giraffe who eats acacia thorns is mainly one of quantity. All three are meat, and a cat can benefit from any one of them if he can catch it. So while the glaciers came and went, while the vegetarians struggled against all odds trying to digest new plants and adapt themselves to overwhelming global changes, the cats simply kept on hunting, waiting to pounce on whoever managed to survive into the next epoch. The limber cat body that hunted successfully in the Pliocene hunts just as successfully today.

Hence, meat-eating has formed cat bodies, beginning in the mouth with daggerlike eyeteeth suitable for fastening their owner to a victim, and with strong, triangular cheek teeth, capable of severing the victim's spine and shearing his flesh into bite-sized chunks for passage down the cat's throat. Meat-eating has caused the shortness of the cat's intestine, since meat is easy to digest and doesn't require a long, heavy gut that would weigh a cat down and keep him from accelerating quickly—a basic requirement for the feline lurk-and-leap style of hunting. Meat-eating explains the short digestive period, the rapid passage of food through the cat, and the nutritional residue in a cat's feces, which is why dogs forage in cats' litter boxes and why cats mark with spray. A spray is not as visible as a scat, perhaps, but at least it will still be there when the owner returns to check on it. Thus, finally, meat-eating even explains why a cat can twist his penis. Like a gardener spraying roses, a cat can direct his urine upward to moisten the undersides of leaves where other cats will find it and where rain won't wash it away.

The most important fact about meat-eating, however, is that it explains a cat's emotions, or some of them. Many expressions of a cat's feelings seem deeply related to the capture of live prey. An excited, happy, or much relieved cat may ambush and pounce upon whatever triggers its pleasure—something worth considering before getting a large cat all worked up.

One moonless night when Lissa was out for the evening, Ruby escaped from her pen. Finding herself at large on the isolated homestead high in the Colorado Rockies, Ruby must have felt anxious and unsure of herself while she waited for Lissa to get home and straighten things out. When Ruby heard Lissa's car, she crouched low beside the house, and as Lissa groped her way through the almost prehistoric darkness that surrounds her mountain fastness, Ruby joyously sprang on her from behind and bore her down. It was her way of expressing relief.

I have occasionally witnessed similar episodes, the most touching involving a tiger, whose version of a free-floating predatory dream was displayed backstage at an outdoor circus one very hot day. The tiger grew increasingly excited at the approach of her trainer, then abruptly stopped leaping and spinning and quickly crouched low to hide behind a solid partition in her cage. When the trainer passed the partition, the tiger sprang at him with her fingers stretched and her claws out. Especially touching, I felt, was the fact that the trainer had not come to feed this tiger but to squirt her with a hose to cool her, and she knew it. The prospect not of food but of cool water, and the joy of playing in it, had fired the tiger's excitement so high that she saw herself leaping from ambush. That her concept of climax was to seize her trainer—as she surely would have done without the bars to stop her—is simply a meat-eater's way.

Conversely, a committed meat-eater may express affection and even gratitude toward his or her prey—a touching and thoroughly appropriate emotion in a creature for whom captured animal protein is the only source of food. Or so that emotion should seem to us, since in many human societies people do exactly the same thing when thanking or venerating an animal who has been killed for food. In a tender scene I happened to witness on the African savannah, a lion and some lionesses were rendering the carcass of a female kudu. The lion took the intact but severed head of the kudu between his paws and, holding it upright so that she faced him, slowly licked her cheeks and eyes intimately and tenderly, as if he were grooming her, as if she were another, beloved lion. Rigor mortis had not yet stiffened her muscles—under his tongue her eyelids opened and shut in a lifelike manner. An infant lion pushed up under his father's elbow and helped to wash the kudu's face.

Even more touching was a scene that aired on public television several years ago. Shot through such a long lens that the image appears flat and blue with distance, the film shows a large male puma who evidently has just killed a large male bighorn sheep. The sheep is lying dead on his left side. The puma lies down full length on his right side, face to face with the sheep, gazes fondly into the sheep's eyes for a moment, then reaches out his paw and tenderly pats the sheep's face as a kitten might pat its mother.

Finally, meat-eating, and meat-eating alone, accounts for a cat's sense of fun, of play. The only forms of amusement ever attempted by any of the cats are simulations of hunting, whether with toys, with each other's extremities, or with live prey. All three of our cats bring live prey into the house, to release and chase, either singly or together. When we see the three cats lined up, peering under a radiator or a bookcase, we know what we'll find there. So, too, do we know what we'll find when we hear creatures rushing about in the dark, banging and thumping. My husband and I have great sympathy with the need of cats to hunt, so except to put the bird feeders very high, we don't interfere with our cats as long as they hunt outside, but it hurts us to watch them torturing their prey, and when a cat comes inside with her head high and her jaws bulging, we rescue her victim.

Considering the great behavioral similarities found among the different kinds of cat, and considering the number of actions that cats appear to perform while on automatic pilot, it seems paradoxical that they show so much individuality. Although every cat lover has seen astonishing deeds done by cats, I offer a few observations of my own, mostly because, as examples of what cats do that cannot be preprogrammed, they seem spectacular. A few cats, for instance, understand their owners' feelings—or to put it differently, many cats may understand their owner's feelings, but a few cats seem to want to affect them. I was told of a cat who, upon finding her owner lying exhausted and weeping on a bed, lay down beside her with her belly curled over her owner's face and, putting her front paws around her owner's head, enfolded her owner as she would have enfolded a kitten.

Cats also appear to understand the use to which human beings put their things. A cat of ours who brings mice and birds into the house has on several occasions put her victims on plates or in bowls. Many cats know what doorknobs are for and can manipulate them successfully. And unlike dogs, who, as everyone knows, are apt to regard toilets as drinking fountains, many cats understand the use to which people put toilets, and a few can even figure out an appropriate technique for using one without falling in.

Housecats are not the only cats with such abilities. Evidently pumas have them too. One night, again at the Denver Museum of Natural History, Lissa was bringing Ruby from her place of confinement to the lecture hall. Ruby tugged so hard toward the ladies' room that Lissa assumed she wanted to drink from a toilet and didn't refuse her. In the restroom, however, Ruby forced her way into a stall, clambered onto the toilet, and, with her four feet on the seat and her tail aloft, she defecated into the bowl. What a cat!

My grandmother, Bessie Merrill, had as a young wife a similar experience with a cat. Much to the delight of members of the family, this cat was often seen using the toilet. In those days, indoor plumbing was relatively new, and a toilet was considered too intimate an object to discuss openly, so the talents of this cat were not celebrated outside the immediate family. If my own cats would do the same, I'd tell the world shamelessly, but they haven't quite got the technique of perching on the edge of the seat, tail high, front paws together, rear paws spread. They know what a toilet is for, though, and they occasionally relieve themselves around the base of the pedestal. This attribute of cats is particularly remarkable in light of the fact that cats in the wild don't seem to have formal dung middens, where everyone in the local population defecates. Cats do spray where other cats have sprayed, or in the same area, but the object of this is almost certainly territorial. Nor is the traditional use the only use to which cats put toilets. Our cats cling to the seat while leaning low to drink from the bowl. But our son's cat dips his left front paw down into the bowl while clinging hard to the seat with the others, then licks the water from his paw. Finished, he jumps down, shakes his paw dry, and walks away.

Finally, cats understand many words of their owners' language, although they often don't seem to. I happened to notice this when my black cat, Wicca, would sleep on top of my word processor. As her body absorbed the heat from the monitor her tail would dangle in front of the screen. "Your tail, Wicca," I'd say, and push it aside. One day I was saying "Your tail" just as the phone rang. Wrong number. When I turned back to clear my view of the screen, I was surprised to see that Wicca had already moved her tail. So the next time her tail dangled, I didn't touch it but just asked. Eyes shut as if fast asleep, Wicca simply moved her tail so that it folded around her body and didn't hide the screen. And from then on, all I ever needed to do was to say, "Your tail."

Wicca, who specialized in hunting birds, died tragically in the claws of a bird, a great horned owl. One of her successors is a cat named Christmas, who also sleeps on my monitor and also dangles her tail. To my surprise, it turned out that Wicca wasn't the only cat who understood the meaning of words. If asked, Christmas sometimes also moves her tail. However, Christmas seems to want something in return for doing me so great a favor. Seizing my hand with both of hers, claws out, she brings it up to her head so that I can scratch her.

Finally, I offer the doings of a cat named Wazo, who lives in Massachusetts with my friend Margie Born. An intelligent, mature male, Wazo is interested in dogs, whom he manipulates. As in many households with dogs and cats, Wazo is fed on top of the refrigerator so that the dogs won't eat his food. When he eats, the dogs eye him from below. From time to time he picks out a kibble, puts it at the edge of the refrigerator's tablelike top, and flips it over with his paw. A dog finds it on the floor and eats it. At one time, Wazo also fed the dogs from bowls of candy that had been left on a table, but as he saw it, candy was fit only for dogs. He himself didn't partake of it, although the dogs ate it wrapper and all.

Yet there is a reciprocal aspect to the relationship. Wazo is a very territorial cat who values his space and hates to see another cat intrude upon it. Nevertheless, in his community, which is urban, many other cats are always prowling about. When one comes into his yard, Wazo goes into the house and somehow manages to communicate the presence of a rival to the dogs, who then burst out the door and chase away the stranger. Margie doesn't know how her cat communicates his problem to the dogs—but communicate he does. In the wink of an eye, the dogs are out of the house and the strange cats are gone.

Most species of cat have never been studied, to be sure, but because cats as notably different as tigers, pumas, and housecats show surprising amounts of individuality, it is probably not unreasonable to guess that the other kinds of cat might too. It is as if Gaia has said to the cats: Here, my beauties—the information you need in order to hunt, mate, fight, yowl, be cautious, raise children, is safely packed in the back of your brains. Save your forebrains for whatever creative inventions may strike you; use your wits to amuse yourselves and me.

S cientists sometimes divide the cat family into two tax-ons, big cats and small cats. The big cats are the lions, tigers, jaguars, leopards, snow leopards, and clouded leopards. The small cats are all the others, twenty-eight species in all. But what is the basis for the division between big and small?

Not size, evidently, since the smallest of the big cats, the clouded leopard, is smaller than the biggest of the small cats, the puma. And not the old adage that big cats can't purr and small cats can't roar. Well, perhaps small cats can't exactly roar, since they haven't got the bodily mass for it, and at the base of their tongues is merely an ordinary tongue bone, the hyoid bone, where big cats have a flexible cartilage, possibly to help produce the deafening vibration. But big cats purr or hum. Leopards certainly purr. And I once heard an interesting story about a lion purring. Admittedly, the story is anecdotal and un-substantiated, and I don't even recall the source, since I heard it over thirty years ago, and at that distance my memory is faulty. Nevertheless I brazenly repeat it anyway. Years ago, in the country that was then the Bechuanaland Protectorate and is now Botswana, a lion is said to have seized a missionary by the nape of his neck and his shoulder and, straddling his body, dragged him into the shade. Before the mission students could drive the lion away and rescue their teacher, the missionary believed he felt an ongoing vibration in the lion's body which he took to be purring. Also interesting was that the missionary reported feeling neither pain nor fear but rather a trancelike detachment that helped him endure this very trying experience. One cannot help but note that cats purr when they are very ill or in pain, and that purring supposedly is related to the release of endorphins. Could the purring vibration in the lion have had that desirable effect on the missionary? Could purring help to soothe other victims? It would help a cat a lot if his victim ceased to struggle and simply accepted fate. But that evolution could promote the reaction seems unlikely to say the least.

Purring is a communication. Enough cannot be said about this fascinating feature of the cat tribe. It is probably the first and most important link between a mother cat and her kittens. They feel the vibration with her warmth as they smell the milk in her breasts, and they grope toward her body. Small cats purr to people probably for the same reason that they purr to their mothers—a reassuring sound that means to the cats what a smile means to people. Interestingly enough, purring affects the human psyche just as smiling seems to affect the feline psyche—our signals, in this case, cross and are mutually understandable. Also very interesting is that no person has yet been able to pinpoint the purr conclusively. A veterinarian and cat specialist, Dr. Richard Thoma, trying to locate a cat's purr with a stethoscope, found that the sound was equally loud all over. Several theories of purring have been advanced—one holding that it is the product of turbulence in the blood stream traveling up the windpipe and into the sinuses. More accepted is the theory that the purr results from a complex interplay involving the voice box and air pressure in the throat. However, another veterinarian, Dr. Richard Jakowski of the Tufts University School of Veterinary Medicine, suggests that the organ of purring may be the soft palate—the thing that shuts when we swallow so that food doesn't get into our lungs. Dr. Jakowski finds that a cat's soft palate is much longer than necessary if its function were merely to separate air from food, and it contains the kind of muscle that is controlled voluntarily. In other words, a cat can will his soft palate to flutter.*

Possibly the reason that the big cats are presumed not to purr is that they tend to reserve the communication for their children. People are seldom present on such occasions and so don't often hear it. And no cat purrs unless someone is around to listen. Those fortunate enough to have heard a lion purring report the sound as a hum. Like the purr of the small cats, the hum or purr of the big cats is produced in the throat, not in the mouth, and seems to be related to the sound of greeting made by some of the big cats when they chuff, or do *prusten*, a sound like a discreet Bronx cheer which a person can replicate by blowing through the lips. As the purring of small cats can be either soft or loud, so the chuffing of big cats can be with or without vocalization—as small cats purr loudly when their feelings are strong or when the person (or cat or other creature) to whom they're purring is relatively far away, so big cats add voice to the chuff for emotional emphasis, if they're really glad to see whoever it is they are greeting. Our cat Rajah, realizes that his beloved owner, my husband, is slightly hard of hearing, and he raises the volume of his purr until he knows my husband can hear him. Even from afar, a listener can tell when the cat and my husband are together because the purring is extra loud.

The main difference between the sociable vocalizations of the big and small cats seems to be that big cats chuff only briefly, whereas small cats purr on and on. The tale of the lion and the missionary is further complicated by the fact that lions don't seem to chuff or purr in greeting but rather give a high, discreet moan. If in fact the lion was purring when he dragged the missionary, was he expressing gratitude to the missionary for what he, the lion, believed he was about to receive? A queer thought, but not at all impossible—like the puma who patted the face of

---

* *Catnip* (Newsletter of the Tufts University School of Veterinary Medicine), vol. 1, no. 9 (December 1993).

his victim, cats of all sizes express their satisfaction at the immediate prospect of food.

One characteristic that seems to distinguish small cats from big cats is the manner of eating. A small cat typically crouches over its food, all four feet neatly on the ground, in the familiar posture of a domestic cat eating from a dish. A big cat, on the other hand, typically lies down to eat and holds its food with its front paws. A small cat often begins eating at the head or neck of its victim, while a big cat usually starts at the haunch or belly. Even so, the form of eating has more to do with the size of the food than with the size of the cat—a tiger given a small piece of meat does not normally lie down for it but simply laps it up off the ground in passing or crouches directly above it, housecat style. Or if such a bit of meat is tossed to him, the tiger may catch it with a snap of his small front teeth—his incisor teeth—and then push the morsel into his mouth with the back of his wrist or with the side of his paw.

All cats tend to pluck some of the fur or feathers before eating. Lions and tigers remove the quills from porcupines, and housecats pluck or partly pluck birds, if only of the tail and pinion feathers. Here, small cats show a further behavioral division: cats with origins in the New World are said to tear the feathers off cleanly, with a strong, upward sweep of the head, while cats of the Old World tear off the feathers with a shaking motion, a motion that rids their lips of feathers and that they begin almost as soon as their lips touch the corpse. Small cats characteristically remove the viscera as well as the hair or feathers, a fact which the owners of free-ranging housecats hardly need to be told, as these viscera are the squashy little things we step on in the dark or find on the floor in the morning. Big cats tend to eat the viscera, but first they cut the silagelike cud from the rumen (if the victim had a rumen) and then they clean the gut by dragging it between their clenched incisors to pop the feces out.

Is this all that distinguishes small cats from big cats—table manners and a hyoid bone in place of cartilage? One might point to a few minor differences, such as the fact that big cats usually sprawl when they lie down, while small cats often tuck their tails around their bodies and fold their paws under their chests. Also the sheathing of the claws is more symmetrical on big cats than on small cats. But these are not major differences.

The fact is, the important thing about big cats and small cats is not that they are different but that they are the same. And like so many other truths about cats, their sameness is due to their diet and their hunting. So perhaps the greatest measure of the cats' success is that the basic cat body, with very little change in shape, comes in so many different colors and sizes. Even the basic call, *meow*, seems to be shared among species, for when a cat wants something (if she sees a bird through the window, for instance) the *meh, meh, meh* of her call is the first part of the *meow*, and when a tiger wants something (another tiger, for instance) the booming *eow, eow* of her call, the moan, is the last part of the *meow*.

A result of the range of feline size has been that, until very recently in evolutionary terms, few if any land animals were too large or too small to escape cat predation. Diminutive cats caught the mice and the butterflies, while giant saber-tooths caught the elephants and the rhinos, with other cats of assorted sizes catching the animals in between. Today, one of the largest carnivores on land is a cat—second only to the polar bears and grizzlies is the Siberian, or Altai, tiger, a huge but silent deer hunter of the ice and snow, who can measure sixteen feet from nose to tail and weigh over eight hundred pounds. Two species of cat could claim to be the smallest—*Felis rubiginosa*, the small, rusty-spotted cat of India and Sri Lanka, and *Felis nigripes*, the even smaller black-footed cat of the African savannah. Adults of either species can measure just fourteen inches from nose to tail and weigh two or three pounds. These cats, of course, can live on worms, insects, and the smallest mice and birds.

However, that they can doesn't mean that they always do. Even these tiny cats can also prey successfully upon creatures much larger than themselves. The black-footed cat has been reported to kill sheep, something I for one did not believe until Rajah, the housecat, began to hunt white-tailed deer. But is this really so surprising? Male American bobcats, who weigh about fifty or sixty pounds, regularly hunt and kill deer that outweigh them by a factor of three. And Daniel Boone and his brother reported seeing a puma killing an American bison, which would have been eight times heavier than its assailant. (Complaining that it was a gruesome sight, the Boone brothers, who themselves were hunting the same bison, shot the puma.)

As cats changed size to better exploit the bountiful smorgasbord of prey animals, they also changed their camouflage and thus can hide or hunt in a wide variety of habitats. The cats of northern climates are often soft colored and grayish, since they must match the background during all four seasons, which is particularly difficult to do in snow. People who work with fabrics know that colors mixed with gray seem to blend.

In contrast to the cats who live in snow, cats of the tropics are often brightly striped or spotted to match the dappled sunlight falling through the leaves. Cats of the savannahs are plain colored, usually tawny, and match the dry grass. Interestingly enough, they often match the green grass too, to the extent that their coats are often the same tone and intensity of color, so that even if the grass is new after a rainstorm the cats vanish into it easily. Some animals are color blind, so would be particularly disadvantaged by the ability of red cats to merge with green grass, but even birds and primates with acute color vision have difficulty in spotting a red object in a green field, or vice versa, if the tone and intensity of the colors are the same.

Even more interesting are the cats that could be considered exceptions—the cheetah, say, who is a savannah dweller if ever there was one but who is spotted nevertheless, and the puma, who has successfully occupied all habitats, from the taiga forests of northern Canada to the pampas of Chile and Argentina, and has spots only in infancy. How so? Probably because, in the case of the cheetahs, camouflage isn't too important, since the cheetah spots his prey from afar and dashes after it. In the case of the puma, since no single camouflage could match all the habitats, perhaps a generic camouflage has emerged. Or perhaps pumas were colored to match their main prey, the deer. Where the deer hide, so do they.

To be unseen and undetectable is a state that seems to suit the cat family. Frequent grooming removes much personal odor from the small cats (although the grooming itself produces a faint odor—an observant friend, the late Susanna Schweitzer, once pointed out that few things smell as delicate, as lovely, as a freshly groomed cat). Furthermore, cats walk so quietly that they are inaudible. In this, they are of course in dramatic contrast to many other animals, including us. We move forward as if by falling, saving ourselves at the last moment by quickly extending a stiffened leg, heel down. As inertia carries the body onward over the extended foot, the rest of the flat sole slaps the ground. Almost instantly we start to rise up on the toe, casting ourselves forward, and repeating the noisy process over again.

Cats walk differently. A cat lifts out each paw like a person offering to shake hands, then gently places the paw, outer edge first, very carefully on the ground before shifting any weight onto it. A very interesting observation on the footsteps of cats was made by the forest ecologist Susan G. Morse, who is also a tracker, and is the only person, as far as I know, ever to notice what is probably the single most important difference between dog tracks and cat tracks.* Most people believe that dog tracks can be distinguished from cat tracks by the claw marks, because cat claws are retractable while dog claws aren't, so that claw marks are said to be diagnostic. But this isn't always so. My husband's dog, for instance, manicures his own toenails by biting them, keeping them so short that on one rather embarrassing occasion his tracks passed briefly as puma tracks. And one of our cats, Silent Spring, for reasons best known to herself, would sometimes walk with all her claws out, leaving tracks that could almost pass for a dog's if they weren't so tiny. A more reliable distinction, Sue Morse suggests, is that a dog's second and third toes are almost perfectly symmetrical, while a cat's paw is like a human hand in that the middle finger is almost always a bit longer than its neighbors. Thus an asymmetrical print not only

---

* Susan G. Morse, personal communication.

distinguishes a cat track from a dog track but also distinguishes the left paw from the right. Yet perhaps the most interesting observation ever made about dog and cat tracks was also made by Sue Morse, and is this: that a dog walking on a soft matrix such as dust, mud, or snow leaves a tiny ridge of the matrix between the toes and the large pad, as if between the tips of the bent fingers and the forward edge of the palm if a similar print were made by a human being. In other words, a dog grasps the earth as he walks, squeezing up some of the matrix if it is soft enough. A cat, in contrast, lays down his foot very smoothly, gently, leaving no mark but the faint dents of his pads and only then if conditions are optimal, such as after a dusting of fresh snow. To me, Sue Morse's observation is as moving as it is insightful and original and shows a basic emotional difference between dogs and cats—that cats tend to hold back, approaching life with reserve and caution, while dogs tend to be a little anxious or eager, always pressing, however slightly, always a little tense.

It also explains why cats walk so quietly, and it even explains why cats, more than any other animal of similar size, often walk without leaving any tracks at all. Thus, many of the wild cats manage to live among us without our being any the wiser. We don't see them, we don't hear them, and we don't find their tracks.

On several occasions when visiting Lissa Gilmour, owner of the puma, Ruby, I searched for tracks and other sign in Ruby's pen. I wanted to find clues to the presence of a puma, clues that I might recognize later. And I did. In the dust of the pen I saw part of a road-killed deer that had been hidden in leaves; farther on I saw puma hairs, puma scats, and a puma scrape; beyond these I saw a well-worn puma trail, and at the end of the trail, on top of a kennel, I saw the puma herself, pretty Ruby, who had been watching my progress with big yellow eyes. But I saw not a single footprint anywhere.

Even the puma herself didn't stay visible for long. When Lissa playfully called out "Yoohoo! Mountain lion!" Ruby crouched low and vanished in seven inches of grass. Quite literally, she was gone. We couldn't see her, at least not for what she was. She had made herself so flat that only the curve of her back showed, like a sand-colored stone or a dry patch of Colorado earth.

For a moment, nothing moved. Then Lissa turned and started running, and suddenly the enormous puma materialized as if from nowhere and came hurtling through the air. In two bounds she was on Lissa. Down they went, with Ruby's mighty paws clutching Lissa around the body and her daggerlike eye-teeth very close to Lissa's head. All this was accomplished in absolute silence. In a moment, Ruby stepped back, averted her eyes from Lissa, and sneezed lightly. Lissa got up and brushed off her clothes. But as the three of us smoothly strolled toward Lissa's parlor, where, in pleasant congeniality, we would soon take tea together, I couldn't help but notice that although this strenuous activity had produced a few scrapes from Lissa's sandals, not a trace could be found of the puma.

So much for the theory that no tracks means no cats. If the experts and authorities say differently, then they cannot be called experts or authorities.

Yet there's more. Where wild cats are hunted, they don't like to leave footprints, which in fact may be the most compelling reason for our failure to find tracks. In African game parks, lions trudge heedlessly along the dusty roads leaving long trails of footprints just as if they were people, but in the New World, pumas, lynxes, and bobcats who commonly use dirt roads as thoroughfares will quite consciously keep to the parts of the track where the dirt is hard packed so footprints won't show. On snowy ground this is particularly dramatic, especially when the cat comes to a patch of snow too big to walk around or to jump over, and can't help leaving a footprint, showing that he's been there. Otherwise, he threads his way around the patches of snow, keeping to the bare earth, even if this forces him to go out of his way.

Why so discreet? The almost perfect discretion of the cats who depend on stealth is part of their camouflage, their secrecy. An animal who understands hunting, either because it hunts or because it is hunted or both, as is the case of the New World cats, is very well served by discretion in all forms. That is why one can live among cats without realizing they are present, a state which was probably as true in the Miocene as it is in Ruby's pen in Colorado, or even as it is in the New England woodlands today.

Most books on cats include a chapter on the cat body, and this book is no exception; however, this book will not repeat the obvious, that cats keep themselves clean, say, and have retractile claws. Rather, here I will focus on two features rarely mentioned in cat literature: the awesome strength and the highly developed senses that are found in every cat alive.

Pound for pound, cats are by far the strongest animals many of us ever encounter. When my husband and I recently tried to wash our cat, Rajah, for example, it took our combined strength just to keep him in the washbasin, and we really needed a third person to put on the soap. All we had sought to do was to rid Rajah of fleas, but both of us got seriously scratched in the process, and the bathroom got drenched from ceiling to floor. In the confusion, poor Rajah escaped, dried himself off, and had nothing to do with us for days thereafter.

When one projects the strength of that seven-pound cat into a seven-hundred-pound tiger, one gets some idea of the animal's might. So it's a good thing that tigers bathe themselves, or so says the science writer Sy Montgomery.

The sense organs, too, are developed to a degree we find

hard to imagine. Cats have six kinds or more, if, as some people feel, psychic powers are also present. Cats have whiskers, teeth, and skin to provide the sense of touch; eyes, ears, and nose for sight, sound, and smell; the tongue for taste; and a nearly invisible system of pits or pores known as the vomeronasal organ in the upper gum and palate for detecting the presence of certain chemicals in the environment. Among other signals picked up by the vomeronasal organ are the estrus pheromones.

Of all the senses, the sixth is least understood. The sea mammals and the higher primates were thought to be the only mammals in which this sense was missing, although some of the higher primates such as ourselves were thought to have it vestigially—the vestige was supposedly the little lump in the roof of the mouth. Or so it seemed until 1993, when a paper was published announcing that the vomeronasal organ in human beings may be functional after all. Exactly what purpose the organ serves for us has yet to be determined.

To use the organ an animal stretches his or her mouth to expose the pores in an act known as "doing flehmen." A tiger doing flehmen grins a terrible grin that stretches out her tongue. A housecat makes a tight little smile, with her front teeth covered by her upper lip, which is stretched tight. Cats often do flehmen, for instance, when they encounter the urine sprays of other cats. Our cat Lilac, mother of the hunting housecat Rajah, slowly and deliberately does flehmen when she encounters a glass of white wine. Other cats scratch dirt over wine and beer, as if covering pools of urine. Why? No one knows. Probably the answer lies in the chemical properties of the alcohol and additives, which are such that most if not all tiger trainers abstain from alcoholic beverages before a training session or a performance. Tigers are said to so deeply dislike the smell of alcohol on someone's breath that, given the opportunity, they will shred the drinker.

Perhaps a cat's most important sense is sight. To understand the impact of eyesight on cats, it is useful to contrast the way dogs and cats respond to films and television. Many dogs seem well aware that what appears to be in the TV set is an illusion and unreal, yet some dogs respond anyway, usually to sound, particularly to the sound of other dogs barking, but sometimes to the image as well. Usually this happens to naive dogs. Once a dog of mine named Koki, a sled dog who began life on the end of a chain in a village in northern Canada, reacted with restrained excitement to the image of a deer in someone's home movie. She leaped eagerly into a chair to better see the screen, but quickly realized that whatever she was looking at was an illusion, and quietly got down. On another occasion, when a television set was placed on the floor, a young female dingo named Viva became very taken with an ad for pet food in which a little team of horses ran through a room and disappeared into a cupboard. Viva tore around the set to catch the tiny horses when they ran out the back. But Viva and Koki were exceptional, probably because of Viva's youth and Koki's rustic origins. As a rule, not much experience is necessary for dogs to ignore the images of television. Koki soon realized her mistake, and young Viva, ashamed, was never fooled again, although I and other loudly laughing people tried everything we could to get her to repeat the mistake.

In contrast, cats seem drawn to the flickering screen. At bedtime, our three cats appear as if by magic to lie at the foot of our bed to watch the evening news like an audience in the front row of a theater. Videos made especially for cats are particularly popular with them. When Rajah was shown a video of fluttering birds made expressly for cats, he couldn't get enough of it—squarely in front of the screen he sat, his eyes huge, his ears in silhouette against the light, his whiskers straining. Every now and then the image would prove too much for him and he'd fling himself against the glass, only to drop to the floor. Reality. At that point, he might leave the room for a while, but soon he would quietly return to the enthralling spectacle. He quickly learned the sound of it, and the twittering of the sound track would draw him from elsewhere in the house. At the time of this writing Rajah still watches the cat video even though he knows it's an illusion, and his excitement is nearly as strong today as it was the first time.

Christmas has also learned to watch this particular video.

Now and then she leaps at the screen, and so involved is she with the images that after leaping she sometimes looks down at herself, backing up as she does—she believes she has caught the bird but has lost it again, and is checking to see if it has escaped underneath her, as mice sometimes do.

Certain people might offer the anecdote of these cats and their video as evidence of cat stupidity. Not so. The anecdote merely illustrates the nature of a cat's hunting instinct. And if cats are proved stupid by having such an instinct, why then, so are we. One of the most important observations I've ever encountered concerned the human hunting instinct and was made by Ken Jafek, a hunter and professional outfitter in Malta, Idaho. It will ring a bell with anyone who loves the natural world. Ken said that the intense excitement that accompanies the sighting of a wild animal does not diminish with repetition or with familiarity or with time. Although Ken is now a grandfather, he gets the same rush when he spots a deer (or other wild animal) as he did when he was a youngster, although in his capacity as a guide and a hunter he by now has encountered wild animals hundreds or even thousands of times. I'm sure this important reflex is found in many of us. It certainly is in me. And here we share something with the cat tribe after all—the stimulus to our hunting instinct in both cases is visual.

A cat's sense of smell, in contrast, seems to have relatively little importance as a hunting tool but is more than useful as a way of learning about other cats. Unlike most hunters, including wolves and human beings, the big cats (those whose habits are known) sometimes ignore the wind when hunting, even though they surely know of their own strong personal odor. Some cats are just as apt to stalk their prey down the wind as up, not always to advantage.

A cat's weak response to airborne scent seems particularly interesting in light of the fact that cats use their sense of smell to find things and to identify other animals. I watched an example of the latter skill at our home in New Hampshire after an elephant and her trainer, who is an acquaintance of ours, had walked past our house. Our cats had been locked up with the dogs in a detached, windowless garage while the elephant was present, and after they were released and were gladly making their way back to the house, they stopped dead when they crossed the elephant's trail. Whoa! They didn't know what the smell was—they had never seen or smelled an elephant—yet they wanted to know! Hair bristling, they lingered long over the footprints, pressing their nostrils to the earth, audibly snuffling. The dogs, in contrast, ran right past the scent trail, much to my surprise. But then I realized that they would have been catching whiffs of the elephant during their imprisonment, so the smell would not have come as a shock.

Their respective reactions were as typical as they were informative. The dogs wanted most of all to join us, their social group, after being kept from us while something as strange as an elephant was present, while the cats cared less about their group and more about their surroundings. They wanted to know what strange thing had visited their home. Furthermore, if it is true that my dogs had caught the scent before the cats did, I take that not as evidence of cats' lesser ability but rather as a sign that they do not give airborne odors the same credit that dogs do. After all, every year at Christmas these same cats use olfaction to detect our family's hidden hoard of Christmas presents, then to find their own presents in the pile, then to remove and open these packages. Our cats are never wrong; they open only their own presents and no others. Obvious? Not really. Why does a cat who can find catnip not search for odors in the wind? I think that cats experience odors as pools, not as trails, as dogs do. A kitten finds his mother by entering the pool of her scent, then finds his own personal nipple by sniffing through the fur of her belly. What this must mean to a cat is suggested, I believe, by the preference cats show for their owners' clothes. To lie on one's owner's sweater, in a cloud of one's owner's scent, must awaken secure and pleasant feelings.

For cats, intraspecies communication is probably the most important function of their sense of smell. Cats have tiny scent glands on their faces—on their lips, chins, cheeks, and at the base of their whiskers—which they rub on us, on each other, and even on objects, mixing their odors with the odors of other cats in a gesture of unity and bonding. Similarly, cats have scent glands on their backs at the roots of their tails with which they perfume other, familiar cats and objects in and around their homes. When a cat arches its back and rubs against you, it is probably marking you with this friendly odor. Cats also have glands between the toes and inside the anus which leave secretions that may carry a hostile message. Cats bury their feces when at home, perhaps to mute the message. But when far from home cats purposely leave scats on top of rocks or hills or other conspicuous, windswept locations for the information of other cats. Thus scent also serves as a Keep Out sign.

The entire face of a cat, including the lips, the fine hairs inside the ears, the points of the eyeteeth, the tips of the whiskers, and perhaps even the surface of the eyes, is an area of extreme sensitivity, and perhaps functions in the same way that facial vision functions for blind people. As a cat walks, his whiskers point forward—forward and down if he's calm, forward and up if he's anxious or excited. A circus tiger often enters the ring wide-eyed and with his whiskers pointed far forward, as if unsure of what will meet him as he emerges from the chute. Usually, the cause of concern is another tiger already in the ring. If the newcomer reaches his seat unmolested, he may feel more confident and will await his first cue calmly, with his whiskers to the side. If his trainer then comes to speak to him, or if he happens to be seated beside a tiger whom he likes and who likes him, he may chuff and bow his head in greeting, with his whiskers flat against his cheeks. But if he is then expected to do a trick that's difficult for him, his whiskers come forward in advance, as he gathers himself to perform. Interestingly enough, human beings can replicate these emotions by moving the skin around the mouth in a comparable manner, pursing our lips when unsure (an act which, if we had whiskers, would push them forward) blandly smiling when serene (to lay the whiskers back against the cheeks).

Using his face and all its senses, a hunting cat keeps his whiskers forward, his lips very slightly parted, his ears up and his eyes very wide. Even his eyeteeth are fitted with nerves, making the tips of the teeth highly sensitive, so that when the cat grabs his prey by the neck and bites down, his eyeteeth feel his victim's vertebrae, finding the spaces between the bones where he can cut through the spinal cord. If the cat is a small cat, he may then shroud his victim with his whiskers to learn through them whether any vibration remains in the victim—in other words, to learn if it's still living. Usually, the small cats don't start eating if the victim is living. So perhaps through his sensitive whiskers a cat learns the moment when eating may begin.

Since all cats are basically so similar, what force produced their thirty-five different species? Probably their habit of roaming. Sometimes we have difficulty imagining how far it was possible to roam during prehistoric times because we wrongly see the earth's dry land as, essentially, two huge islands, with the New World floating on one side of the planet and the Old World floating on the other. This image is of European origin, whereby the Atlantic looms large. And so it should. The first land mammals ever to cross the Atlantic were the Vikings.

The Pacific is much bigger, but the Pacific could sometimes be crossed if one knew where to find the bridges and the stepping stones. We would picture the earth better, therefore, if we saw the earth's dry land as a single continent, like a horseshoe bent around the planet. If we stood at one end of this continent—on the coast of Scotland, say—and instead of looking west over the Atlantic and wondering how to sail across it we turned our backs on the ocean and began to trudge east, our biggest problem would be to cross the English Channel, which on occasion has been little more than a stream. Once on the east side of it, we would tramp onward through Europe and Asia.

Eventually we would reach the Bering land bridge, which may be under water today but not so long ago was a cold and windswept tundra and may be so again. Crossing the land bridge, we would wander eastward across Alaska and Canada into Newfoundland, where we would again reach the Atlantic. And once there we would have to stop.

Near the beginning of the voyage, we could have made a lengthy side trip down into Africa, and near the end of the voyage we could have made a similar trip into South America. Thus, most of the earth's dry land was accessible to creatures who roamed over it on foot, at least during certain phases of its history. And roam is exactly what the cats did, not just once or twice but countless times, traveling in all directions.

Like their distant relatives the modern lynxes, the lynxlike Ur-cats, *Pseudaelurus*, probably lived in forests. No one can say for certain what they ate, but one can easily imagine them hunting large rodents and ground-nesting birds. However, the pressures of different habitats were working on the Ur-cat's body, and eventually the animal began changing to better exploit the various parts of its range. In South America about ten million years ago, the ocelots branched from the parent stem and, along with other little spotted mousers of the New World jungles, became known as the ocelot lineage.

Meanwhile in the Old World, wildcats appeared. Colored to blend into the shadowy forests, they were tabbies, with a large, long-haired form, *Felis sylvestris*, for the forests of Europe, and a smaller, short-haired form, *Felis sylvestris lybica*, for the forests of North Africa.

The forests of North Africa? Yes, North Africa was verdant until the Neolithic, when people domesticated many of the ungulates, invented the concept of wealth, and began to display their status and their social importance with the size of their herds. The Book of Job records, for instance, the generous compensatory damages awarded to Job for his suffering—23,000 head of livestock, far more than Job would ever need. With all these animals, Job must have been one of the richest and most powerful pastoralists of his time or indeed of any time, a man of great importance, overgrazing with the best. According to Scriptures,* only 13 percent of Job's animals, the cattle and donkeys, were water dependent, while 87 percent, the sheep and camels, were desert tolerant, suggesting that whoever organized the gift seemed to have known how best to exploit a dry and overgrazed pasture. The preference for sheep and camels further suggests that livestock had been chewing up the suffering landscape long before Job's animals put their lips to it. The result was desert, from the Negev to the Sahara.

Still camouflaged for the nonexistent forests, the African wildcats stayed on after desertification, hunted mice in the granaries of North Africa and Asia Minor, and became housecats, *Felis sylvestris catus*. Thus, just as we originated in Africa, so did our cats, and today the wildcats and their domestic cousins are known as the domestic-cat lineage.

Meanwhile, something must account for Siamese cats, Himalayan cats, Burmese cats, and their relations. So, undoubtedly, the Asian version of the wildcat, *Felis sylvestris ornata*, had not been idle. Perhaps penetrating the granaries of India and China, the gracile cousins of the African cats seem to have ingratiated themselves with the human population there as well. Unlike the Egyptian cats, they didn't gain civil rights or get themselves deified into an Asian version of the goddess Bastis. Even so, they became the treasured pets of monarchs and to this day, for whatever reason, are often accepted by fanciers as a higher order of cat. This, of course, is a human concept and has absolutely nothing whatever to do with excellence or competence, which both the Asian and African versions of the housecat seem to have in equal shares, depending not on the ancestry but on the individual.

Another big step in cat evolution was the rise of the pantherines. There is still some disagreement as to how the ancestral forms relate to one another, but at least one version puts the pantherines in the New World with a wildcat ancestor, proba-

*Job: 42:12.

bly a wildcat who had padded across the land bridge in fits and starts. This ancestor took its time in becoming a true pantherine, with many experiments along the way. Perhaps the simplest experiment was to grow very large. Not too large, though—the saber-tooths were already the big cats of the New World, preying on the big herbivores and probably living in prides. Better to stay mid-sized in the face of such competition, big enough to hunt deer when circumstances allowed but small enough to hunt rabbits when the deer were gone. The result was the puma, really nothing but a small cat grown to large proportions, who is still counted as a small cat and who keeps a small cat's ways. Pumas purr loud and long like housecats, for instance, and crouch above their food. Soon, pumas had spread over the entire New World, from the estuaries of Maine to the coastal forests of the Pacific, from the peaks of the Rockies to the swamps of Louisiana, from the evergreen taiga of Canada's far north to the open savannahs of the Argentine.

Few animals are as capable or resourceful as pumas or have been as successful. Even today, after having been exterminated throughout much of their former range, pumas are returning in eastern Canada and New England, where their habits seem to differ somewhat from the habits of western pumas in that they are even more shy. Nevertheless, they are here and are seen more and more often.

Was the ancestor of the pumas also the ancestor of the cheetahs? The two cats have much in common. Like pumas, cheetahs roamed the New World. Also like pumas, the cheetahs preyed on deer and other deer-sized ungulates and in hard times could, like pumas, turn to rabbit-sized prey. But the cheetahs hunted by coursing their prey on the open plains, whereas the pumas hunted by stealth in long grass and in thick forests. Thus these cousins avoided competition and could live together in the New World.

Perhaps because cheetahs are thought of as African and pumas as American, the two are seldom compared. Even so, powerful similarities exist. For instance, pumas and cheetahs sound alike; both make trills and chirps like birdcalls and sharp, clear whistles that sound like a person whistling with two fingers in the mouth. In fact, these whistles are made with the voice and are extremely thin, high-pitched *meows* used for communication between mothers and kittens or between pumas or cheetahs and people or other creatures with whom the cats have a positive relationship. The thinness of the sound ensures that it doesn't travel far, which helps to keep the whereabouts of the loved one or loved ones a secret.

Both cheetahs and pumas have pupils that stay round and don't form slits like the pupils of wildcats and housecats. Both cheetahs and pumas have long hind legs that tilt their hindquarters upward and give them a graceful, ankling walk. Both cats have long, heavy tails to give them balance. Both cats can move almost faster than the eye can see, although cheetahs sustain the movement far longer and are said to be the fastest mammals on earth. Yet it is hard to imagine any animal much faster than a puma—in support of which I offer an observation of Ruby, Lissa Gilmour's puma, on Lissa's handsome carpet, amusing herself one quiet afternoon. Seemingly half asleep with boredom, Ruby lay on her side while her yellow eyes followed the circling of a fly. Suddenly her forepaw made a blur and the fly vanished. She had snatched him from the air, not with her fingers but with her palm. Idly, she turned her paw over, palm up. Clasped in a single crease that had appeared across her leathery pad, a crease that a palmist might have called her lifeline, the fly kicked his tiny legs, struggling.

Soon Ruby's paw relaxed. The fly turned over and soared away, unharmed. Ruby watched him without seeming to watch him, as if the fly might be frightened by her stare. When he came near, her paw blurred again, and again she had him. What other animal could do the same?

Both cheetahs and pumas rely on long-distance vision for much of their information, and both use lookouts—high places from which to survey the passing scene. Both cats give birth to three or four kittens at a time, usually in a hidden nest or a cave in the rocks, and both cats breed at any time of year. Year-round breeding may not seem so unusual in cheetahs, since they live in warm climates where many other animals also breed year round, but it seems astonishing in pumas, who are rare if not unique among indigenous wild North American mammals in having such an ability.

The fate of pumas and cheetahs diverged somewhat when cheetahs crossed the Bering land bridge into Asia and became a global species, ankling along to India, to the Middle East, to Africa, and to Europe. Farther, no cat could go.

Why didn't the equally successful pumas follow? The answer probably lies in meat, as usual, and, if so, also explains why tigers didn't come the other way. These cats probably couldn't have crossed the land bridge—a tundra without trees or cover—because being stealth hunters they almost certainly couldn't have hunted there. Plenty of animals would have grazed on the tundra, but these were migratory animals such as bison, elk, and reindeer (caribou), who would have visited the land bridge in high summer. Like reindeer today, the migratory herds of long ago probably used the windblown snowfields as sanctuaries from the Arctic's terrible mosquitoes and biting flies. Coursers such as the cheetahs and group hunters such as the cave lions (and probably also the saber-tooths) could have fed themselves on the open tundra, just as lions and cheetahs feed themselves successfully in broad daylight on the open plains of Africa. Smaller cats such as the wildcat who gave rise to the puma could also hunt on open tundra—even short grass hides the smaller cats. And, indeed, the wildcats and the lynxes crossed the land bridge many times. But tigers and pumas need good hiding places, and on the exposed plains of the land bridge these larger cats would have been too visible. On similar open plains farther south in America and Asia, the pumas and tigers could have caught their prey at night, but on the land bridge in high summer there was no night. Darkness didn't come until autumn, but by then the migratory herds would have been on their way to treeline, to their sheltered, winter ranges. In contrast, the little voles and lemmings that fed the small cats didn't migrate seasonally. Rather, they sheltered in grass tussocks or in burrows under the snow and thus provided food for wandering wildcats and lynxes all through the year.

So perhaps the most interesting members of the pantherine lineage are the lynxes, heirs to the cat family's ancestral size
and habits. Numerous lynx types have appeared over time, some
in Africa, others in Europe, still others in Asia. One in particular, an Asiatic lynx, spread both east and west, the westward travelers arriving in Europe to meet a previously established
European lynx, and the eastward travelers arriving in Canada
to encroach upon another established lynx, an earlier immigrant, the bobcat.

Today in Canada and the northern United States the two
kinds of lynx live side by side. The more recent immigrant,
known as the Canada lynx, is adapted for the far northern snows
with huge feet and heavy fur frosted with white (perhaps because he had come more recently over the Bering tundra and
simply stayed in country he had grown used to); the earlier arrival, the bobcat, has faintly dappled spots, which give good
camouflage in wooded, brushy habitats from Canada to Mex-
ico. Thus have these two cats of similar size and appearance divided the lynx-size feline's share of the New World.

Invariably, two similar animals dividing a habitat find ways
to diverge, so that they become less rather than more alike. The
two kinds of lynx were no exception, but interestingly, one point
on which they diverged seems to have been personality. Why
this is so, no one can say. Yet Canada lynxes are said to be calm,
even gentle, while bobcats are supposedly spitfires. Perhaps
their different life-styles suggest the reason—the main prey of
many lynxes is the snowshoe hare, for which they are particularly well adapted, while many bobcats must hunt deer (or so
they have evidently been doing, at least in the northeast, since
the rabbit population declined). Probably only large male bobcats are capable of killing deer, which must be extremely difficult even for them and must put a premium on courage and
aggression, or at least on the ability to overcome fear and charge
right into the task at hand. Perhaps the ferocity of bobcats is
the by-product of centuries of hardship. No one knows, of
course. And very few people care. The studies we choose to fi-
nance on the subject of bobcats relate mainly to their role as

furbearers so that vain and selfish women can parade around wearing scraps of hide torn from these capable animals. Other than facts that help us trap them, we know so relatively little about the daily lives of cats in the wild that we have few mechanisms even for approaching the more complex questions that could be asked.

The lynx family gave rise to the so-called big cats: the lions, tigers, ounces, leopards, and jaguars of the genus *Panthera*. Like the lynxes, all the big cats are spotted, at least for part of their lives, except for the tigers, who are born striped and whose coats, unlike the coats of lions, say, don't change from childhood to adulthood. But are tigers truly striped? Maybe not. A tiger's stripes are actually elongated spots, a secret that is betrayed by certain individual tigers whose stripes are conspicuously open in the center, so that each stripe is shown for what it is: a leopard's rosette greatly elongated, like a stretched rubber band.

Beginning about two million years ago with the appearance of the clouded leopard (bearing the beautiful name of *Neofelis nebulosa*), the big cats evolved in increasingly larger sizes to hunt increasingly larger prey. Today, the biggest animals hunted by cats are probably wild cattle such as the African buffalo and the Asiatic gaur, with the cats dividing these populations so that tigers get at them in the forests and lions get at them on the plains.

Most of the big cats were highly successful as colonizers. In the Old World, the leopards were widespread, with the ounce

(*Panthera uncia*) and the leopard (*P. pardus*) in much the same relationship to each other as the Canada lynx is to the bobcat—the ounce adapted for snow-covered mountains, and the many races or subspecies of the leopard adapted for the jungle and bush. In the New World the jaguar managed to colonize most forested parts of the Americas except in the far north, thus competing for the deer-sized prey with its older and more gracile cousin, the puma.

Few animals, however, have ever been as successful as the lions. The most recent of the *Panthera*, lions evolved on the African savannahs only 700,000 years ago, probably to the consternation of our ancestors, and soon thereafter—long before our ancestors thought of doing anything similar—the lions colonized the rest of the planet. Until very recently lions were almost everywhere that glaciers were not. They lived in North and South America, all across Asia and down into India; they lived in Europe and on the British Isles; they lived in the Middle East and throughout the continent of Africa, where of course they remain to this day. (Australia and the Pacific Islands were about the only places the lions hadn't found.)

In keeping with the tendency of animals to be very large in the northern parts of their ranges, the northernmost lions were enormous. These were the cave lions. Having survived the Ice Age, their hair would have been longer than the hair of their southern relatives. They may have been quite fluffy. Only recently extinct, they certainly coexisted with people in Europe, Asia, and the Americas and may have been observed by at least one Chinese painter, who made a scroll showing a huge, fluffy cat—tigerish but not a tiger—whose pale fur was plain except for some stripes around its tail and knees. Perhaps this painting shows a cave lion. If so, it would not be the only cave lion image left by people, by any means. In some Magdelinean sites in Europe, Paleolithic artists carved large bas-reliefs of lions to lurk in the darkest recesses of the caves.

Today we think of tigers as Asian and lions as African, but this is a misconception; the two species coexisted in Asia from the

Caucasus to China and from Korea to Iran. A few Asiatic lions still live in zoos, and a tiny relic population remains in Gir, one of India's national parks, where each one of them knows and is known by each of the park rangers personally.

Lions and tigers are so much alike in size and shape that it is difficult to distinguish their skeletal remains. They can even mate and produce young (the tions and ligers bred for zoos, circuses, and private menageries). That the two cats don't interbreed in nature is due in part to the habitats they seek and even to the times of day they sometimes favor, the tigers tending to be nocturnal forest dwellers and the lions, if perhaps to a lesser extent, diurnal plains dwellers.

By the middle of the Pleistocene, cats of all sizes lived all over the world except in Australia. However, the cat was so well designed and so successful that the Australian marsupials created a cat of their own. Called *Thylacoleo carnifex*, or "marsupial lion executioner" (literally "meat-maker"), the animal evolved, touchingly enough, from possums. These possums, in making their own version of a lynxlike creature, adopted the cat family's most popular size—small enough to subsist on mouselike animals but big enough to tackle deerlike forms. But while a true cat's longest teeth are the eyeteeth, this possum-lion's longest teeth were the front teeth, or incisors. His threats would have looked like a squirrel's, or like Dracula's. *T. carnifex* supposedly went extinct long ago but nevertheless is believed by a handful of optimists to exist to this day, and in that extremely unlikely event would be clinging to life somewhere in the far reaches of the Outback.

Meanwhile, as the true cats spread themselves over the earth, they managed, for the most part, to adjust so perfectly to one another by keeping their distance, by hunting different prey, and by moving around at different times of day that now different kinds of cats can coexist on the same bit of ground. Three square miles of national forest in Idaho might, for instance, support a puma, a bobcat or two, and perhaps a few feral housecats, with the puma hunting deer and the bobcats and housecats hunting birds and small rodents and keeping well out of the puma's way. A few square miles of tropical rain forest might similarly support a jaguar, a puma, an ocelot or two, and one or more varieties of the smaller spotted cats. In a European forest, to the consternation of roaming housecats, a lynx's range might also support some wildcats, and in the Asian jungles, a tiger's range would probably support a leopard and possibly also one or more kinds of the small, rare, and beautiful jungle cats. Finally, the range of a pride of lions on the African savannah could support up to five other feline species, most of them in slightly different settings. One might find a black-footed wildcat, a serval (Africa's version of the ocelot), and perhaps a caracal (Africa's version of the lynx). Larger cats might live there, too, perhaps a leopard to hunt in the long grass and thickets and a cheetah to course the open plains. Such cats are competitive and mutually hostile; they seek and destroy one another's kittens but otherwise avoid one another at all costs. Even so, by overlapping the distribution of their species, the cats have so perfectly arranged themselves that virtually everything that creeps or flies or walks can be used by one or another of them in almost every situation. Or at least, that's how it was before habitat destruction, the pet trade, and the fur trade took a heavy toll on the cats.

Cats are well known for their homing instincts. At a recent Christmas party in our town, one of the guests boasted that he had successfully rid his family of their housecat by driving her to Concord, the state capital, some thirty miles away, and tossing her out of the car. Shocked, other guests asked him why he had done it, wondering, I'm sure, why he hadn't taken the cat to our local humane society, which costs nothing, does not euthanize its animals, and is much nearer than Concord. But the man misunderstood the question. He had been forced to go all the way to Concord, he explained, because he had already tried abandoning his cat at shorter distances, and each time she reappeared.

The cat hasn't returned from Concord, at least not as far as is known, suggesting that the owner's enormous and callous cruelty was greater than his cat's ability to home. Meanwhile, the disturbing episode raises a question: Why did the cat keep trying to return to those people?

The answer is that many cats care more about their homes than their owners. Housecats feel for their homes strongly and obviously, but other kinds of cat do too, and many cats, both housecats and wild species, have made some very impressive journeys using navigational skills that are hard to explain. A radio-collared puma, for example, whom researchers had transported in a closed box by plane for relocation in a new environment, evidently didn't like his new surroundings and headed for home, arriving there four months later after a journey of three hundred miles.*

Why is the pull of home so powerful? Again, the answer lies in meat. Every cat must live amid a population of prey animals, whether mice or bison, in a place where there is suitable feed to keep the prey from wandering away. Every cat also needs water, and every cat needs shade or shelter, particularly if she is female, since she will need several hidden nests in case she must move her kittens. To meet these requirements, each cat tries to establish and hold a territory which, of course, varies in size according to the species of the cat and the habitat of the prey animals. The size of a territory also varies by the sex of its owner. Female cats of most species tend to own adjacent or almost adjacent territories of relatively modest size. These fan out through a jungle or drape down a mountainside, sometimes overlapping very slightly, sometimes touching at the corners like the petals of a wild rose. A male cat tends to own a larger territory that may partly overlap the territories of several females or may encompass the territories of these females. If their territories are the petals, his is the rose.

When a cat finds a place with the necessary requirements, he or she marks out the boundary with scratches, perhaps placing a few scats on high places, and squirting out a great many sprays. Some if not all species of cat even have two types of spray for the purpose, one like the jet from a water pistol, the other like the mist from a plant fogger. At least to a human nose the two kinds of spray seem different, one being more odoriferous, so perhaps the cat mixes the secretions of a gland into its urine when spraying but withholds the secretion when merely relieving itself.

* Maurice Hornocker, "Learning to Live with Mountain Lions," *National Geographic*, vol. 182, no. 1 (July 1992), p. 52.

Hence, except for the fact that most cats tend to change their territories rather often, the boundaries of cats are for all the world like those of human beings, although set in claw marks and urine rather than in fences and surveyor's tapes. Yet in both cases, the boundaries have a psychological effect on members of the same species. Not many people can genuinely ignore a No Trespassing sign. Even if they trespass, few do so without some twinge of conscience and most must overcome a certain amount of reluctance. Cats are the same. Coming upon the sprayed urine of a property owner, a cat will stop and sniff and grimace, and perhaps back off to think for a while. It may then go forward, but only after having made up its mind.

Thus cat properties are like ranches. The space enclosed by the cat's boundaries is actually the grazing land for livestock, whether deer or deer mice, which belong to the owner and to no one else, and which the owner does not disturb except to harvest.

When new land is difficult to find, mother cats often share with their children, keeping their sons with them until they are full grown and keeping their daughters or some of their daughters even longer, sometimes even dividing the ranch with one or more daughters. Lions and housecats carry the practice to an extreme, so that the members of a pride—really a group of females who are related to each other, grandmothers, mothers, daughters, sisters, nieces, and grandchildren—own one enormous ranch together. The area owned by a pride of lionesses can be hundreds of square miles, particularly in southern Africa, where game is sparse, while the area owned by a group of housecats is usually someone's farm, particularly the barn and its surroundings, where most of the mice are. Female cats share nests with their sisters and help to birth, groom, feed, protect, and teach one another's children.

Cheetahs are something of an exception to the paradigm of cats as ranchers, since female cheetahs seem not to own land. Male cheetahs do, though, and form coalitions, sometimes of brothers, but sometimes just friends, who together own a territory. From time to time a female cheetah joins them on it. If the males are successful, they defend the territory against takeovers by other males and offer the female a safe place to raise her children.

Thus, if the need to eat meat explains the cat body, so the need for a ranch on which to raise the meat explains cat society. The social organizations of cats are so different from ours that we seldom recognize them for what they are. Even so, they are there, very important to the cats, who with some exceptions merely practice different variations of the family theme. To us, lions in a pride on a savannah seem to be organized very differently indeed from solitary tigers in a forest, and yet from a cat's point of view the organization of the two species could be very much the same. Imagine, for instance, a limp balloon with spots painted on it to represent a pride of lions. When the balloon is inflated, the same spots could represent a community of tigers or other so-called solitary cats. The distance between group members might be different, but the relationships of the individuals might be the same.

Because cats are otherwise so much alike in their needs and habits, their social similarities are perhaps not surprising. For example, contrary to what many people believe, tigers not only have recognizable social units but also sometimes live together in small groups, usually consisting of a mother and two or three of her full-sized but still subadult young. But even mature male tigers sometimes join these groups. A 1989 newsletter of the International Union for the Conservation of Nature and Natural Resources (IUCN) reports that in Kanha Park in India the filmmaker Belinda Wright observed the meeting of a mature male tiger with a mother tiger, her two small cubs, and her full-grown son from an earlier litter. The five animals greeted each other with enthusiasm and affection and eventually departed in a group.

Therefore, that all cats don't live in prides may have little to do with their social inclinations but much to do with the type of food available to them. A tiger, after all, is roughly the same size as a lion and eats about the same amount at a sitting. Yet in many localities the prey available to the two kinds of cat differs greatly in size. Out on the savannahs, any one of the large African antelopes provides a meal for many lions together and may encourage their cooperative hunting and sociability. But deer and pigs are the mainstay in many tiger habitats. Scattered in dense forests and too small to feed many tigers together, the potential prey may encourage tiger solitude.

Still, the mystery remains. What explains the apparent paradox of areas where lions are found living singly at least part of the time even where large game is plentiful? The question may have to do with water and will be considered further on. Like many other questions about animals, the problem is as fascinating as it is complex.

Is the concept of cats owning property far-fetched? I don't think so. All wild cats need territory—that's why they establish home ranges. But where does the concept end? This question, I think, has no simple answer but should not be ignored. Why own a home range if no food can be found on it? I submit that the cat who owns the property also has a proprietary interest in the food supply. In fact, to keep the food source for himself or herself is the single most important reason for the cat to claim the property in the first place. In this spirit, housecats display a proprietary interest in their owners, marking them with wipes of the lips, rearing up to incise their legs with claw scratches, occasionally spraying them and their belongings and defending them from the advances of other cats. Why? Because a supply of food issues from a housecat's owner in much the same way that a supply of wildebeests issues from parts of the savannah or a supply of Arctic hares issues from a few square miles of taiga forest.

A human paradigm can be found in hunter-gatherers—the Ju/wa Bushmen, for instance—whose sense of land tenure very closely resembles that of certain other species and for exactly the same reasons: people must have access to food and water and must protect these commodities from takeover by other people. The ability to control access to one's food supply is life itself. This is so obvious it scarcely bears mentioning, and is as true of any other territorial species as it is of us.

If asked, a group of Ju/wa Bushmen might not say in so many words that they owned a stretch of land. And they don't own land in the way we own land. They don't buy it or sell it or transfer ownership in any way, since land is not a commodity. Rather, they own land like we own air. As we might say we had the right to defend the purity of our air so that we could all breathe safely, a group of Ju/wasi might say that they had a right to hunt the game in a certain area, to gather the wild vegetables, and to drink the water, and that any newcomers who might wish to do the same would have to ask permission. "For everything?" you might ask.

"Yes, that would be right," the Ju/wasi might answer.

"Even to hunt hyenas?" you might ask. "Do I need permission for that?"

"Oh, no, not for hyenas," the Ju/wasi could say. "Hyenas are *chi dole*, worthless things. We wouldn't care. You could kill as many hyenas as you wanted. It's amusing that you would think to ask permission for that!"

So hyenas aren't part of the deal, you say to yourself. "Elands, then," you ask. "And kudus. What about my hunting elands and kudus?"

"Ah well—elands and kudus," answer the Ju/wasi. "It is better to ask permission before you start killing the game. Kudus and elands are useful animals, hoofed animals. People care about them."

In this scenario, then, did the Bushmen *own* the elands and the kudus? Not according to our standards, to be sure, by which someone could buy or sell or even rent those animals. But the Bushmen owned them by another standard, by the standard of land tenure and land stewardship, by the standard of a territory as a place complete with flora and fauna where one could live without unwholesome competition from others of one's species, and where one stood ready to dispatch other predatory species, nuisance competitors of about one's own size such as cheetahs, leopards, and hyenas. In short, ownership as the Bushmen experienced it was of an ancient standard that held as long as the owners lived in the old way. In this way the cats, too, own land and the game animals upon it. The lionesses of a pride own, say, the wildebeests on their territory for as long as the wildebeests stay.

Ownership of a ranch, whether communal or individual, often makes the difference of life or death to a cat. Not only does the ranch provide the cat with shelter, food, and water but the land and its inhabitants become familiar to the cat in great detail, which is extremely helpful to any hunter. Cats often rest where they can view their holdings—they watch the movements of their livestock and can assess the condition of each animal they see. In all likelihood, they even know some of the prey animals personally, so they know who is aging, who is sick, who has calved or has been bested in a battle or weakened by ticks. Such knowledge combined with knowledge of the cover and topography cannot help but contribute to successful hunting. Once in an Idaho forest I was shown the carcass of a deer killed by a female puma who obviously had known that a deer trail led through a certain heavy growth of trees and had reached the area by creeping through a shallow draw. Chance could hardly have brought her there just at the exact moment when a deer was passing. Rather, she had obviously planned the hunt in advance, had crept along where she knew she could hide, and had perhaps been waiting for the deer. After killing it, she had eaten her fill, and then, covering the carcass with leaves to hide it from birds, she had gone off to rest at a little distance. But she had kept her food under surveillance and had returned to eat from it again and again until all the meat was gone. We found only the shins with hooves and scraps of hairy hide, the parts that pumas usually discard when preparing their food before eating it. A year later, these parts still lay near the edge of the draw as a tribute to an excellent hunter, a Diana of the douglas firs who understood not only her prey but also the topography.

In contrast was another Idaho puma, a young female with four kittens, perhaps her first litter, who seemed homeless. I was able to briefly trace her fortunes with the help of Ken Jafek, whose expertise and whose marvelous hounds enable an ongoing puma study. Ken had found evidence of this young puma a few days before I joined him. Because the young female had not previously been noted by the study, when Ken first found her tracks near an old, abandoned farm he felt sure that she had come from afar. At her heels trailed her four children, only three months old, still fluffy and spotted. Why had she chosen the coldest part of winter for her travels? Why had she come to that barren valley where no deer stayed? No puma who realized that the place offered so little game would have chosen it for a new home. So surely her presence there meant that in her former home something had gone wrong. She had chosen the valley because she didn't know where else to go.

Still, a stream ran through the valley with thickets growing along the banks, so the puma and her children had drinking water and small amounts of food in the form of hares and other little creatures. Soon, Ken found that she had managed to kill a porcupine, which she and her children had eaten, leaving nothing but quills on the scraped and frozen skin. By the time I arrived and was shown the scraps, the young mother had covered them with snow, as if to save them for later.

Very soon after her arrival, the researchers set a dog to chase her up a tree so that they could fit her with a radio collar and monitor her doings. But this panicked stranger didn't climb a tree. Instead, unfortunately, she kept running, followed by her kittens. The little family ran right through a grove of fir trees that would have saved them and up an open, snowy slope where the dog overtook them and caught one of the kittens. Nothing could be done—the kitten was dead before the researchers, far behind, could intervene.

That night it snowed, and the scattered family had regrouped but had not left the valley. Instead, the mother had brought her surviving children to an old, long-abandoned farmhouse that stood crumbling beside a little road. When Ken and I came looking for this puma in the predawn hours of an icy morning, we found her tracks in the snow, showing how the mother and her children had hurried toward the house, not in single file, as they normally would travel, but four abreast, as if all had welcomed the sight of it.

Inside, they took shelter from the bitter night under the broken boards of the living room floor. We tried to approach quietly in hopes of seeing the pumas, but the mother heard us. Once again she fled with her kittens behind her, out the back door as we approached the front. The departing tracks told us that here the kittens had run behind their mother, following in her footprints.

Wanting to identify them for the study, Ken loosed a dog— not an experienced old warrior like the dog who had killed the kitten but a tender little pup only a few months older than his quarry. Nose to earth, and amazingly professional for one so young, the puppy quartered the willows by the stream, his tail waving briefly with each encouraging odor. This time, two of the kittens hurried up a tree, and the third, a female, climbed into a bush that didn't seem high enough or strong enough to keep her aloft very long. But the youthful dog barked "Treed!" and we came scrambling. The mother was long gone.

We spent the day there, tying the dogs to trees and sending for the researchers in Pocatello, who arrived hours later, to tranquilize, measure, and radio-collar two of the kittens before letting them go. (They would have collared the third as well but had brought only two collars.) I thought, from the changing orientation of the tied-up dogs, that the mother might be circling, and on her behalf I tried to keep the kittens warm and safe until the drug wore off. They weighed about twelve pounds apiece, which spoke well for their mother's care-giving ability, but they seemed surprisingly light, almost cloudlike, and I couldn't quite make out their edges under their long, fine, soft fur. Yet it was the paws of these kittens that I found most surprising. Huge, like soup plates, these paws seemed out of proportion to the little bodies. And when I gently pressed the velvet toes, great daggerlike, horn-colored claws eased out, strong and curved like an eagle's talons. One, caught in my parka, slit the heavy fabric like a razor in silk. These twenty great weapons on each kitten are a promise of the cat to come.

When the kittens were themselves again we put them safely

into a tree and left, revisiting the place in the morning to learn how they had fared during the night. The mother must have watched our departure from the bushes—she had come for her kittens almost immediately and had led them far away. Judging from the direction she had taken, she was heading back to the place she had come from.

We stopped at the abandoned house to see what we could learn from it and found the fur, heads, paws, and paunches of two small raccoons that the pumas had eaten, not much food, perhaps, but lots better than nothing. We also found the print of the young mother's body in snow that had drifted through the open door. She had lain in the doorway keeping watch through the night, as any parent would do.

This fragment of observation suggests, I think, how disadvantaged an animal is when in unfamiliar country, let alone when in a place where food and comfort are in short supply. Mule deer were plentiful about two miles distant—we later saw a distinguished herd of them, fifty or more, making their way up the side of a high hill. But the young mother didn't seem to know about these deer. Nor did she know where to take shelter or where to hide. Ken Jafek has loosed his dogs upon dozens of pumas (exclusively for scientific research in recent years), and he tells of pumas evading the dogs by such tactics as climbing rock faces, showing that they knew the countryside and all its refuges—in other words, those pumas were at home.* But the young female wasn't at home. A stranger in a strange land, she had fled aimlessly when the dogs first chased her, and after her second experience with a dog, she had moved elsewhere, surely to continue her search for a place to live and raise her remaining children.

Her story is typical of many if not most of the wild cats to-

---

* Hunting pumas or, for that matter, any of the mid-sized cats with dogs is by and large a safe procedure. Except in rare instances, a cat pursued by dogs merely climbs a tree, and the dogs bark up at him from below. If the human hunter doesn't shoot at or otherwise injure the puma, or chase him out of the tree so that he must confront the dogs, he simply waits until the excitement is over, then climbs down and goes on about his business, and no harm is done.

day, whose dwelling places are thoughtlessly destroyed or invaded by people. When we think of cats or indeed of any animals, we comfortably assume that they belong in the woods, and if that is where we find them, why then, they must be in good shape. Yet even the animals that eat grass and leaves have very real, even rigid requirements, so that one place is not just the same as another but depends for suitability on many factors, including the food supply. Just as some parts of a city are vastly different from other parts, so that a New Yorker in Sutton Place would experience life differently from a New Yorker in the South Bronx, so the natural world is not monolithic. Most human beings, especially those in Western and Asian cultures, are so isolated from the natural world that they lack all feeling for these truths.

When I was in my late teens and early twenties, I was privileged to be a member of a series of anthropological expeditions to the Ju/wa and /Gwi Bushmen who lived in a vast, largely unexplored part of southern Africa. I went there with my parents, Laurence and Lorna Marshall, and my brother, John, to record the Bushmen's way of life. While traveling in one of the most remote parts of the desert, we met a little group of ten people—two men, three women, four boys, and a baby—who, like the young puma and her children, had no suitable home. As Bushmen, these people were hunters and gatherers, deriving their entire livelihood and all their possessions from the natural world just as our ancestors had done long ago and just as all wild animals today continue to do. During the dry season, in the way of the savannah from time immemorial, the people would go to live near permanent waterholes, and during the rainy season when water was generally available they would move out over the veldt to take advantage of the foods that grew there. The Bushman people had distributed themselves fairly evenly in those days, and evidently had been doing so for a very long time, so that each waterhole and the surrounding land did not become overcrowded with more people than the land could feed. People therefore experienced hunger no more and no less than the other creatures. They would suffer during prolonged droughts, of course, but they escaped the frightful devastation of famines that result in other parts of the world today. In contrast to what the so-called civilized people erroneously believed about the Bushmen, who were wrongly described as nomads and wrongly assumed to wander aimlessly over the veldt in search of food, they, no less than the other creatures of the savannah, had territorial claims which apportioned the area's resources among their population in a way that gave everyone the necessities of life.

Living in the old way, by the old rules, the /Gwi people in those days experienced life very differently than we do now. Their cultural experience was much more like the experience of the other creatures with whom we still share the earth. Therefore, when the group of ten people lost their home, they suddenly found themselves cut off from their dry-season water supply and in a desperate situation. How had they lost their home? They lost it just as many animals do—their holding happened to be on land that had been taken over by a European farmer. Naturally, he didn't recognize their prior ownership; after all, they had never farmed the land and had no deed or other written claim to it.

At first he didn't need exclusive use of the waterhole, and he tolerated the presence of the /Gwi family just as we might tolerate some foxes, say, or other wild things whose ways didn't particularly interest us. But when he decided that he needed the water for his own use, he drove the people off just as we might drain a swamp or forget to fill a bird feeder, never dreaming that such a deed would cause anyone any harm. After all, the people were Bushmen who lived wild, in the bush. Why should he, a busy farmer, concern himself with them? Why couldn't they disappear into the bush and go on about their business?

What happened to the /Gwi family was exactly what happens to many other creatures who lose their homes—they did not go to some other place to live because there were no suitable places available to them in the dry season. They knew that, but they tried anyway, departing for a place they formerly had used only in the rainy season, relying on their people's greatest skill and art to draw liquid from some unexpected sources in the desert. When we met them deep in the interior in the middle of the dry season, they had no water at all but were doing without it. Yet for all their knowledge and skill, life in the driest part of the desert in the hottest, driest season eventually proved too precarious even for them and they returned to their old place. Still the farmer didn't let them stay, so their family disintegrated. They went where they could, good luck did not follow them, and a few years later, only three were still living. A statistical look at survivors from other mammals in similar straits would have predicted who these three would be: the young men. The women, the old people, and the young children perished. This situation has occurred more times than anyone could ever imagine, both to animals and to human hunter-gatherers, and is exactly what had happened and was still happening to the little puma with her surviving, cloudlike children.

Although cats are no less social than many other species, we tend to think of them as solitaries, probably because whenever we see a cat he is physically far enough away from the other members of its group as to seem alone. Celebrating the concept of cats as solitaries, Rudyard Kipling described the cat who walked by himself, waving his wild tail in the wet, wild woods. In fact, however, even when spread out on their ranches, cats are highly social. If they weren't, they wouldn't need the vast repertoire of vocalizations, scent glands, or tail and body postures to convey an almost unlimited number of emotional impressions to other cats. Nor would they need their highly mobile facial skin—their delicately muscled ears, eyelids, lips, cheeks, and whiskers—to produce a wide variety of changing facial expressions. The faces of cats are more mobile even than dog faces or our faces. The unsocial animals, by contrast, don't need facial expressions and hence their faces seem stiff.

My husband and I saw the importance of facial expression in cats dramatically illustrated by our cats after one of them, our beautiful white Aasa, went blind. We had no idea that anything was the matter with this cat. Her blindness was caused by an inherited condition that we were unaware of and that had come on gradually. Because Aasa had known our house and its surrounding land from childhood, she managed to get around almost perfectly even after she became blind, so that the only changes we noticed were that she seemed very quiet and subdued, that she had developed a strange way of getting onto a bed—never jumping but reaching stealthily with her paw and feeling the surface before clambering up—and that there were fewer little corpses of mice and birds being brought into the house.

But the other cats noticed big differences. Perhaps the most important was that Aasa's response to them wasn't right and wasn't the same. Even to us, her gesture of feeling a surface before jumping up to it seemed eerie, disconcerting, perhaps because she seemed to do it stealthily, without looking at us and without changing her facial expression even when we were right there watching her. Undoubtedly, her failure to do the right thing at the right time provoked the other cats very deeply and caused them to persecute her. They seemed to be especially hard on her when we were away from home. We would return to find no sign of Aasa except piles of white fur that the other cats had torn from her skin. After long searches, we would find her hiding, usually in some dark corner of the basement. We would call her, and she would come part way toward us, moving very slowly, very carefully, particularly when crossing a part of the crawlspace that contained a deep, open well. Looking back, picturing where she would stand and how she would move in the crawlspace, I can see that she remembered the well and was being careful to stay far away from it.

Eventually we would catch her and carry her upstairs. Sometimes the three other cats would then chase her, and in flight she would often narrowly miss a doorway and run into the wall. Even then we didn't get it. We knew that something was terribly wrong, but we had no idea what. Before long, we had to section off the house, putting Aasa in one part and the rest of the cats in the other. But still the other cats stalked her, peering at her through an interior screen door. She had no way of knowing they were looking at her and therefore failed to respond to them appropriately, which to them was inflammatory behavior and only made them try to get at her all the more. Because we didn't know what the matter was, we weren't able to observe what mistakes she made that brought on the attacks, but my guess is that because she was a junior female in our household's cat group, she didn't realize when a senior cat was looking at her. She couldn't meet the other cat's eyes, nor could she avert her own eyes, a very important social requirement for lower-ranking cats. Nor could she solicit affectionate rubbing—another feline gaffe. The other cats must have felt a cat's version of the idea that Aasa was deliberately trying to frustrate them, something that a normal cat would never do.

At last, we decided to find her a new home. There, her blindness was discovered by the wonderful, thoughtful people who adopted her. They are Peter Schweitzer and his late wife, Susanna, cat owners extraordinaire. So the strange persecution was finally explained. Aasa now knows her way around Peter's house and is on good terms with an adolescent cat who also lives there and, being younger than the blind cat as well as her junior in the household, is in no position to take offense at any possible mistakes.

Why don't we recognize cat society? Probably because, at least superficially, it differs so greatly from ours. In contrast, we recognize caste in dogs because we rank ourselves by the familiar dog system, a ladderlike social arrangement wherein one individual outranks all others, the next outranks all but the first, and so on down the hierarchy. But the cat system is more like a wheel, with a high-ranking cat on the hub and the others arranged around the rim, all reluctantly acknowledging the superiority of the despot but not necessarily measuring themselves against each other. Farm cats and cats of other closely knit groups usually seem egalitarian. But it is my impression that they may actually have a cat's version of the social ladder, not like that of dogs and people with one individual per rung but rather with several cats on each rung. Even then, the signs of hierarchy in cats are displayed so subtly that our poor powers of observation can barely detect them. A cat expresses soft, submissive feelings by lightly rubbing his face on the faces of those whom he would please: his feline superiors or dogs or his owner—anyone close who he feels is not a rival and who is above him socially. Likewise, a vertical tail can be a sign of superiority in housecats, as is a curved "scorpion" tail in leopards, tigers, and lions. It is important to note that a raised tail in a cat is not the same as a raised tail in a dog, because in dogs the raised tail means higher rank, and thus when two dogs meet, Dog One raises his tail while Dog Two holds his tail low. When two cats meet, however, both may raise their tails, with Cat One raising his higher, perhaps because he feels more confidence. My cats usually go out into the fields with their tails low and usually come home with their tails high—they are coming home to other cats, as well as to the dogs and human members of the family.

In a delightful encounter in 1993 in Harvard Yard in Cambridge, a certain cat who had been hunting birds in the bushes noticed me watching him, gave up his hunt, backed out of the bushes, and purposefully strode right toward me, lifting his tail to the sky and his gaze to mine as he drew near. As our eyes met, he *meowed* once and kept on going. *You have disturbed me so I'm leaving this place*, his gesture said. *But since you are a human being, I'll greet you anyway. Hi there.* Sometimes, though, the tail shows nothing. Two cats simply eye each other and both know who is who.

If the social arrangement of cats differs from ours, so does their social behavior. Here again, human beings are more like the dog family, whose social inclinations we have no trouble recognizing. As an example of the difference I offer my own three dogs and three cats, who each morning choose to follow me to my office in a detached garage. I don't ask them to come with me and I don't feed them when I get there; they are free to come and go via many dog doors. In short, I offer no incentive other than my company. When I start for my office, the dogs crowd out the kitchen door with me, squeezing past my legs and almost upsetting me, never bothering with a nearby dog door because in the morning they are feeling sociable and want to be close. Meanwhile, nothing is seen of the cats. However, by the time the dogs and I are halfway to my office we notice that the cats are also outdoors and heading in the same direction. On we go, the dogs and I exchanging looks, words, and touches, while the cats apparently ignore us and each other. But soon after the dogs and I are inside the office the cats materialize there, too, just as interested in being together as the dogs are but expressing themselves in a different way.

So it is my impression that cats seem unsocial to us only because we aren't good at recognizing the signals of other species. We interpret cat signals to tell us, for instance, that they don't care about us and don't miss us when we're gone. If people were giving off similar signals, we'd be right. But they're not people, and we're wrong. When my husband and I take a trip, we leave our three cats in the house with plenty of food and water. We take the dogs to a kennel, but the cats may not know that, because we drop the dogs off on the way to the airport. As far as the cats are concerned, we and the dogs depart together, leaving the cats at home. And evidently the cats mind this.

No wonder. The cats depend on the people and the dogs, whether they show it or not. When thunderstorms gather, for example, the cats come indoors to be together with me or my husband. If neither of us is at home, the cats stay with the dogs, preferably with my husband's dog, Sundog, who is Dog One in our family. Even though Sundog doesn't like the cats, they look up to him, believing that when cosmic troubles threaten, he'll know what to do. Thus they feel truly abandoned when we go away, taking Sundog and the others but leaving them, the cats, alone in the house. As we leave they retire to the upstairs rooms to lie on the beds, tails tucked and paws neatly folded, where they become unresponsive, a demeanor that resembles human pouting or sulking and may be an emotional withdrawal from an unpleasant situation. If called, they won't answer and won't come.

But when we return they have heard the car and are waiting at the door, tails high, to curl themselves around our feet and lift themselves to brush lips with the dogs (but not always with Sundog, who keeps them at a distance by showing them the tip of an eyetooth and watching them out of the corner of his eye).

I once watched a group of four male cheetahs crossing an African plain, not one behind the other in the way of most land mammals but scattered like crows about fifty feet apart, each finding his own way. These cheetahs had formed an alliance to own a tract of land, and therefore it seemed to me—enclosed as I am in primate sensibility—that they ought to glance at each other every once in a while. In a similar situation we primates would glance at each other, perhaps to see if our companions were in agreement as to where to go next, or perhaps merely to reaffirm our bonds. Dogs, too, would glance at each other. But cheetahs evidently don't feel the need. With their chins high and their half-shut eyes looking at the hills, the woods, the sky, the horizon, anywhere but at each other, they slouched along in bland oblivion. My first impression, stupid as it now seems, was that none of them realized that the others were there. When as if by chance they all reached the edge of the plain together, each chose his own path as they sauntered off among the trees.

As for Kipling's cat, being of the same cattish mindset as the cheetahs, he probably never thought he was walking by himself. On the contrary, he may have thought he was walking with Kipling and, waving his tail aloft as a beacon, was probably trying to lead the author out of the wet, wild woods.

Perhaps the most dramatic evidence of cat sociability is their vocalizations—a logical means for animals whose economic needs drive them apart even as their emotional needs draw them together. People who have both dogs and cats can verify the statement: when called, the common response of dogs is to come, and of cats is to answer.

Lions certainly answer. During a period in the 1950s when I was in Nyae Nyae, in the Kalahari Desert of southern Africa, the lions kept in touch by roaring. One would roar, and after a short while another lion, very far away, would reply. On certain nights the lions would spread out through the bush in a line perhaps a mile long or even longer and seemed to keep their line straight and in order by answering in turn. The farthest would roar, then the next and the next, until six or eight had made themselves known. This way they could tell if all were present and if their line was reasonably straight. In the rainy season, these lions even answered thunder. I loved that: a dark night, the endless, rain-soaked bush, a flash of lightning, a cosmic crash of thunder, a little pause, and then, faint and far, a lion's roar!

"WHERE ARE YOU, MY LION?"

"... (*me?*) ... I'M HERE!"

Among all mammals, and indeed among most other vertebrates, the basic social unit is a mother and her children. But the so-called social animals form strong and fundamental alliances far in excess of the bare-bones maternal tie. Among dogs the unit is the pack, classically a mated pair with their young from several litters. Among people the unit is the extended family, a hunter-gatherer band. But among cats, at least among the species whose ways are known, the unit is the polygamous marriage.

Lions practice a form of polygamy that is also practiced by people in, say, parts of North Africa and the Middle East. In it, a man and his wives and children might occupy a common residence. Such a family would commingle on a daily basis and could travel together as a group. These same conditions could just as well apply to a pride of lions.

The so-called solitary cats are fully as polygamous as lions but with different living arrangements. Theirs resemble, say, those of certain families in Idaho and Utah, wherein a man provides a separate residence for each of several wives and their children and then moves himself from one household to the other, living with each wife in turn. Thus a male puma, say, or a male tiger maintains a large range that he patrols to guard against incursions by other males, a range that overlaps or adjoins a number of smaller ranges that belong to his wives.

Once, in the mountains of southern Idaho, as a short-term participant in a telemetric study of pumas, I found it very heartening and also moving to compare a family we were monitoring on the shoulder of a mountainside with a family in the valley below. The family on the mountainside belonged to a polygamous puma. He himself was probably not present during our period of study, but two of his wives were there with five nearly adult youngsters; one female had twin sons about a year old, and the other female had eight-month-old triplets—a daughter and two sons.

The family in the valley belonged to a polygamous man who had set his homestead right in the middle of the national forest. He, too, was evidently not present, but members of his family were there inside the widely scattered cabins and trailers—the dwellings of his wives.

All day the seven pumas above us slept near one another (actually the two mothers were a quarter of a mile apart with the youngsters scattered between them, which is sardinelike compression for pumas); all day their radio collars were silent— evidently the cats were resting in the shade behind some large boulders. All day the people below us stayed indoors, each woman with her children in her cabin or trailer. So well did each group mind its own business and keep to its daily routine that the only time we became aware of any of them, other than from empirical knowledge, was when a formation of F-111 jet fighters from a nearby airbase came zooming so low overhead and with such a roar that I, for one, thought I was being sucked up by the wake. The two polygamous families also reacted to the disturbance—all the startled pumas immediately raised their heads—not that we could see them, but the action set their radio collars signaling—and several people ran out of the cabins to look up. Of course by then the jets were gone, so the pumas let their weary heads sink down again and the people went back inside. Nothing else moved for the rest of the day except the sun, which slowly crept down to the horizon.

But when darkness gathered and stars came out, the people and the pumas changed activities. Several of the people, who had spent the day apart, came out of their cabins, got into a pickup, and drove away together, while the pumas, who had spent the day together, got up and moved apart. Heading west, the mother of the triplets followed a contour into the next valley. Heading south, the mother of the twins climbed the ridge and vanished over the top. And as each mother chose her own separate hunt, each was trailed at a distance by her children. Big as they were, they hadn't yet perfected the skills of hunting, and they padded youthfully after their mothers, knowing that the mothers would provide.

· Since we saw none of the pumas in this instance, but knew them only by the faint beeping of the radio collars around their necks, how did we know the configuration of their relationships or even who they were? In truth, except for the mother-child and sibling relationships, we didn't know for certain. Bits of tissue, snipped from the poor creatures at the time of their collaring, had been sent to the National Institutes of Health in Bethesda, Maryland, for a study of the DNA that would provide clues to their relatedness, but the researchers were still waiting for the results. One time, the senior scientist on the project told me that the two females were probably mother and daughter, but another time he told me that he guessed their ages, which he would have tried to determine when collaring them, to be about the same. He said, however, that mother and daughter seemed the most likely, and I think most people would agree. The study gave no indication of a male puma in the picture.

Even so, I thought a big male puma was around somewhere, or had been until recently. Why? Because I thought that his image, his shadow, lay over the other pumas. Here, two female pumas lived virtually side by side, yet the kittens of one were three or four months older than the kittens of the other. Since male cats are infanticidal and kill kittens who are not their own, the father of the second, younger litter, the triplets, was probably the father of the twins as well and also was probably the dominant male puma of the area. As to his whereabouts, he could have been elsewhere on his range. After all, the range of a male puma can be ten times bigger than the range of a female puma, so perhaps he was at the far side of his holdings, perhaps on patrol to check his perimeters, perhaps to visit other wives.

But that he was somewhere in the picture seemed likely, or at least, it did to me. There seem to be few other possibilities. If the father of the younger kittens had been a newcomer to the area, he would, in all likelihood, have sought out and killed the older kittens as soon as he arrived.

Why would a cat do something like that? Because he would be forced to. What Gaia expects of male cats makes a dramatic, complex story that begins with the fact that the various cats, unlike many other animals, have two different reproductive strategies, one for the males and another for the females. The social organization brings the strategies into harmony.

Unlike people—or unlike Americans, anyway—and unlike many other animals such as geese, young adult cats of the opposite sex do not meet, pair off, start a family, and hope to live happily ever after. Male cats are more like the Year Kings of Eleusis, who after a period of wandering would supposedly arrive at a land dominated by women, under the reign of a temporary king. A newcomer would fight the resident king, and if he won, he would take over. His reward for victory would be the crown and the favors of the women, yet his fate would surely be that of his predecessor, so no king could expect to remain with the women for long. One wonders what beneficial genetic effect the Year King system might have had on the population of Eleusis, or at least upon its royal family. The practice has certainly contributed to the strength of the cat family, since virtually every male that comes along gets tested by the resident, and the better of the two gets to father the next crop of children. By making themselves available to the victor, the females constantly upgrade their children's genes.

The price of cat excellence is therefore paid largely by the males, and the life of most male cats is difficult indeed. The best beginning a male cat can have is to stay with his mother as long as possible so that she can provide him with sufficient nourishment during his growth and teach him all she knows. As long as he's in his mother's home, and as long as the resident male is his own father, he's likely to stay put until he's physically and mentally ready to seek his fortune. In some cases, his own father tolerates him for a surprising length of time, especially if the youngster lives discreetly, doesn't challenge his father, doesn't advertise himself by yowling (if he's a small cat) or by roaring (if he's a big cat), and doesn't spray or mark. But sooner or later his young male heart prompts him to do some of those forbidden things anyway, and when he does, his father forces him to leave.

Off he goes, to roam where Fortune takes him, in search of females and a place of his own. Now comes the difficulty. His species, after all, has been populating the earth for a million years or more (700,000 years if he's a lion), and his kind has had plenty of time to fill every habitat available to them. Cats have been known to go to breathtaking extremes to overcome this difficulty—a housecat in New York City, for instance, bravely made his way along a hazardous trail of narrow ledges on tall buildings in order to look through a sixth-floor window at a certain female cat named Bubastis. (Sad to say, when Bubastis noticed him looking in at her, she became a bottle brush and spat at him.) Perhaps on the vertical surfaces of skyscrapers in Manhattan, a cat could establish a territory without having to face down too many challengers, but under most circumstances, if the population of the male cat's species is healthy, all suitable homes are occupied. Each is patrolled by a big, experienced tom, a tom who has put scratches on the trees higher than our young male can reach at present, and whose spray drips off the undersides of an astonishing number of leaves. The young male learns that no home is open to him.

So for a while he lives at the edges of other cats' territories, keeping to the parts they use the least, where their scratches are blurred, their feces dry, and their sprays of urine stale. After he has been surprised by one or two of them, perhaps after he has been mauled a few times, he learns to be even more careful. He tries to leave no tracks, he buries his feces, he urinates discreetly downward into holes he has dug for the purpose, and he never, ever sprays. Of course he must keep on hunting, otherwise he will starve. But he doesn't stride around the countryside looking everywhere for animals as his father used to do. Instead, he learns to hunt secretly, crouching patiently near mouse holes (if he's a small cat) or by game trails or in the reeds around waterholes (if he's a big cat), waiting for his prey to come to him. That way, the cat or cats who own the property and whose game he is poaching are less likely to become aware of him. He knows all about them, of course. After all, they spend a lot of their time patrolling, checking the perimeters, constantly on guard against intruders such as he. He may even see one of them go by, since they walk openly and boldly and don't need to hide like he does, at least not from other cats. At night he hears them yowling (if they're small cats) or roaring (if they're big cats). For a while, he manages to elude them. Possibly they suspect his presence, but they can't prove it. Sooner or later, though, he yowls or sprays, too, or is glimpsed by one of them, or otherwise makes a mistake so that they're sure he's around. They then get all excited and hunt him down.

Because cats have such dangerous teeth and claws, they can inflict enormous damage on each other in a fight, and consequently they are inclined not to fight, even in a territorial encounter. Instead, they put their facial expressions and bodily postures to use, each conveying the emotions stirred by the sight of the other. Often, this is enough to end the challenge and frighten the newcomer, who runs away as fast and as inconspicuously as possible—ears flat, tail and belly low to the ground—never to return but rather to try his luck at poaching on other cats' land.

One day, however, our youngster may find a piece of land without a male owner and claim it as his own. Or while he is roaming around or living as a squatter, he may notice, say, that the marks of the owner are not as high on the trees as his marks, suggesting that the resident cat is shorter than he. In this case, he may well seek a confrontation, and because on this occasion both cats have a great deal at stake, a fight may result. How so? Because the young challenger, seeing his chance at a future, has everything to win by fighting, a state fully as compelling as that of the established cat who will fight because he has everything to lose, including his life and his children.

So let us say the two males fight and the young cat wins. Slowly starting to lick his wounds, the young hero keeps his round eyes on the departing form of his bleeding, defeated rival. And then he begins his hunt for the old cat's wives. These may be mothers and daughters, and (if the cats are lynxes, say, or jaguars or ocelots or bobcats or tigers or pumas) their lands may adjoin each other in the cat family's wild-rose configuration. One after another, the young male will locate these holdings and find the nesting places if he can. He has plans for the females, but his plans do not include the kittens of his defeated rival. He is searching for kittens, and when he finds them, he kills them, often with bites to the back of the neck.

By this time, the females know all about him. Any female with kittens fears him. She won't leave her kittens alone any longer than she must, and she fights if the new male finds her nest. Some females manage to drive him off. Most male cats wouldn't fight too ferociously on such occasions—if one observes housecats during male takeovers, one gets the feeling that the marauding male prefers to back off and try later, when the mother cat is absent, rather than to engage in battle and risk injury. Anyway, he undoubtedly wants her goodwill later and would thus prefer to play a waiting game rather than confront her. So despite his greater size and superior strength, the first few times he tries to get to her kittens she usually manages to drive him off.

Still, she knows he'll be back. So if the kittens are small enough, she almost invariably moves them, carefully carrying them one by one to a new hiding place. Most nesting cats keep several possible den sites in mind against just such an eventuality. But when she leaves for the new place dangling a kitten, she may return to find the others limp and still, their spinal cords severed in the neck. And while she bends over them, licking them, trying to revive them, the male may be backtracking her to her new nest to kill the last survivor. Touchingly, when threatened by a tomcat, even a very small kitten seems to understand what is happening and tries to resist, staring and hissing at the marauder.

If the kittens are old enough at the time of a takeover, the mother may try to lead them far away. Her own home range may not be big enough to give sufficient distance from the killer—she may strike out for places unknown. This is not an unrealistic effort, since kittens are amazingly strong. Ken Jafek discovered the trail of a female puma who for reasons unknown had decided to move her three six-week-old kittens through deep snow to a place more than three miles away, a move which by rights should have been impossible for the kittens: the snow was over their heads and too heavy for them to plough through. Yet they completed the journey. True, their mother had done some of the heavy work for them by breaking trail, but their legs were much shorter than hers, so their strides were shorter than her strides. Her footprints were the size of saucers and their footprints were the size of silver dollars. Even so, as Ken's photographs show, the kittens had followed all the way in the trough made by their mother's body, bounding from one deep footprint up over the snow and down into the next footprint; up, over, and down; up, over, and down; repeating this exhausting maneuver no fewer than five thousand times, or once for each yard-long stride their mother had taken.

If kittens are adolescent at the time of a takeover they may strike out on their own. Feeling the menace, the young sons of the defeated ruler may simply decamp as fast as possible. Some of the young daughters, on the other hand, may try to stay with their mothers, but if they do, their fate may depend on how their new stepfather sees them. If he sees them as wives, he won't disturb them. But if he sees them as children, then they, too, may be in danger. George Schaller in his book *The Serengetti Lion* tells of a mysterious observation made by G. Dove that in my admittedly inexperienced view might pertain to this phenomenon. Near Lake Lagaja, Dove watched a lioness approach a feeding lion. At first she walked, but when she got near, she crawled, as if she were begging for the lion's good will. The lion got up from his meal, attacked the lioness and, after a brief scuffle, killed her with a bite to the back of the neck. He then returned to his food. When he was finished he walked over to her and lay down to rest companionably beside her still body.

As Schaller himself has pointed out: "An observer seldom knows the history of an animal he sees, yet its behavior and the responses of others toward it are often influenced by what has happened in the past." Dove's haunting story makes one wonder if the lioness could have been a subadult, a daughter of the pride, trying to appease a new stepfather.*

How long can a male cat control his territory and protect the kittens that, shortly after his conquest, he fathers on the bereaved females? This, of course, depends on him. If he is to make his genetic contribution, however, he must stay long enough not only to father some kittens but also to defend them from takeovers by other males at least until the youngsters can fend for themselves, which is, very roughly speaking, about nine months to a year for most of the small cats, perhaps fourteen to eighteen months for mid-sized cats, and at least two years for large cats. Of course this does not mean that the youngsters are full grown so quickly—lions and tigers, for instance, take three or four years to reach maturity—it just means that some of the youngsters may be able to support themselves if conditions are favorable and if the necessity arises—a minimum age, so to speak. Hopefully, a resident male will be able to dominate an area long enough so that his wives can raise his children to an age when they will surely be able to manage on their own, so that his daughters can settle on or near their mothers' ranches and his sons can go voyaging as he once went voyaging, to start their own families and perpetuate his lineage. Sometimes a res-

* George Schaller, *The Serengetti Lion* (Chicago: University of Chicago Press, 1972), pp. 55, 189.

ident male can dominate an area long enough for his wives to raise more than one litter. However, like the Year Kings, the average male cat will probably reign relatively briefly, with the length of his sojourn depending on his intelligence and strength and his skill at fending off challengers. And if he is not killed when a younger, stronger stranger finally defeats him, he again becomes a vagabond, a voyager, camping temporarily on the holdings of other cats, poaching their game, trying to escape their notice.

This should result in a large population of homeless male cats who are past their prime, and yet no such population seems to exist. For the relatively few species of cat who have been studied, the old male cats who are homeless may be fewer than ordinary attrition would suggest. On the other hand, homeless youngsters are numerous. Surely a possible explanation for at least some of the discrepancy is the amount of fighting the older cats are forced to do, against not one challenger but many. The battles are exhausting, and since cats are such dangerous animals, severe wounds invariably result. Yet in contrast to a young up-and-coming cat who, if he finds himself outgunned, can retreat to fight later at a time of his choosing, the old landowner must put up a full-blown defense against each and every serious challenger, ready or not, and can never indulge in the luxury of retreat. In my opinion, two possible reasons for the apparent scarcity of older, landless male cats may be that they tend to get killed during successful takeovers and that those who survive the final battle are made more vulnerable to ordinary hazards by exhaustion, depression, and their wounds.

Once I was privileged to examine the skull of a saber-tooth. A relic of the La Brea tarpits near Los Angeles, the skull had belonged to a male *Smilodon fatalis*, one of the largest cats the world has ever seen. The skull's mass alone was impressive, but the famous eyeteeth were awesome—strong, curved, pointed, and almost six inches long. The eyes had been large and the chin small, which must have made even this great face somewhat catlike. Beyond that, though, the bones of his lower jaw were fairly narrow, and the bony processes at the corners of his jaw were tiny, just two points, really, so that nothing stopped the jaw from dropping straight down like a trap door.

Other cats have lesser gapes, the result of heavier jaws with wider hinges to anchor the muscles that clench the teeth. But the smilodon didn't need to clench his teeth. What he needed from his lower jaw was that it drop out of the way to let him deploy his sabers. A number of reconstructions* suggest how he might have done this, and although these reconstructions differ on how he mounted his attack, all agree that he drove his eyeteeth home not by pushing up against them with his lower jaw as other cats do but by plunging them down into his victim, driving them home with the vicelike force of his neck and shoulders and indeed with the whole curl of his body. Considering the size and strength of that body and the length of those teeth, to be bitten by a smilodon must have been a terrible experience. If he had attacked a person, and he could have, since in those days our species shared the New World with his species, his teeth could have speared right through his victim like toothpicks through a little pork sausage. Rightly was he named *fatalis!*

While thinking such thoughts, I had been looking at the skull from the top and sides and had, for a few moments, postponed looking into the face straight on. I wanted to be ready for the experience because I expected to get a chill. I expected to look into the face of Death, into the gaping mouth and empty eye sockets of the Great Devourer! But to my surprise, when I put my palms under the sabers to lift and turn the heavy skull, the sense of awe was cancelled immediately by the points of the teeth, which almost pierced my hands.

What had we here? These teeth were perfect, even after ten thousand years in tar! No dings or dents or furrows marred them as they could mar, say, the teeth of a mature African lion. Here was not the Face of Doom—here was the face of a young male cat who had met with a tragic and fatal accident. And since

---

* In Andrew Kitchener, *The Natural History of the Wild Cats* (Ithaca: Comstock, 1991), pp. 29–34.

cats don't grow from kittens without plenty of help, two or three years of his mother's care and effort had been necessary to raise this smilodon to his present giant size. Her warm thighs had sheltered his little body from the glacial winds. The backs of her long heels moving through the grass had been his beacon as he stumbled along behind her. Why had her efforts gone for nothing, and how had a fine young creature like himself ever gotten into a tar pit?

One can't know, of course, but judging from a census of the corpses found together with this youngster, the tar was a magnet to smilodons. With the possible exception of dire wolves, more smilodons fell in than any other animal. Why this happened is not perfectly clear. My own guess is that a tar pit is exactly the sort of peril that would challenge young smilodons off on their own, away from their mothers for the first time, and prone to accidents. Their huge young bodies would have needed lots of food. Small-sized prey would not have nourished them. To get enough to eat without, as it were, overspending their energy budgets, young smilodons would have had to tackle fairly large animals. Certainly, large animals were what they would have learned to hunt from watching their mothers. Even so, hunting very large animals isn't so easily done, especially by a neophyte hunter. Surely the difficulty increased as the populations of the bigger North American mammals dwindled toward extinction—an event that occurred before the disappearance of the smilodons. Furthermore, because smilodons are believed to have lived in prides like lions, a male takeover could have resulted in the cubs of more than one litter suddenly finding themselves on their own in rather close quarters, giving one another plenty of local competition for the easily captured prey. How tempting it must have been to a hungry youngster to come upon an animal evidently standing in a shallow lake, obviously in distress and crying or struggling!

What happens to female cats at times of takeovers? How do they feel about the challenges to their resident husband? It is my impression that female cats have mixed feelings, as is shown by their actions. Among housecats, at least, but surely among some other kinds of cat, too, the females sometimes pitch right in and fight beside their husband when defending against a marauding male, particularly if he comes near their nests and kittens.

But if the newcomer keeps his distance, perhaps staying at the periphery but calling to the females, they may go out to learn more about him and eventually reach familiar terms. If he is a barn cat, he may set up a territory at the edge of the woods, at some distance from the main cat scene at the farm, and lure a female or two over to see him, perhaps even to mate with him. This the females often do quite willingly, an act which is in their best interests in the event that the newcomer should mount a successful assault on the resident male and take over. If the newcomer has been enjoying sex with the local females, he will remember them after his victory. And if his new, promiscuous wives have long been giving him reason to think that their kittens are also his, his interests are served by sparing these kittens when he finds them.

In my early years I lived on a farm where barn cats flourished and where male takeovers must have been reasonably common. The basement of the farmhouse was accessible to the outdoors through gaps in the fieldstone foundation, and through these gaps cats would come to wait for skim milk from the milk separator when, every night, we cleaned and sterilized the machinery. The following is a reminiscence, not an observation, since it happened so long ago. I remember one evening after hours that we heard someone singing in the basement. A strange male cat had arrived and was calling. Never before had I heard such a hauntingly beautiful song. Because he sang for a long time, I learned his song and could sing it to myself for years thereafter.

The door to the cellar stairs was open a crack, and one of the female tabbies slipped down to see him. I followed her and in the dim light saw one of the other females, a tricolor, crouching under the water tank, looking out with round eyes. The strange male was hiding and fell silent when I came. But when I left he began his song again in the dark basement of the farmhouse. A cat Orpheus.

For a while after that, we kept glimpsing the marmalade stranger at the edges of the fields and in the shadows of the outbuildings, and two months later, the tabby and the tricolor females had litters containing marmalade kittens who lived to grow up under the nose of the resident despot, a black cat with a white chin, a white shirtwaist, and four white socks, a cat who in fact looked like the famous Socks, the presidential cat at the time of this writing, but was battle-scarred and more massive. Who had fathered the marmalade kittens? Marmalade coloring is the male version of tricolor and was so common that it could not be diagnostic of paternity, so no one knew. But in a way, actual paternity may not have mattered. What mattered was that neither of the male cats killed the kittens, since each had consorted with the mothers and may have felt that the kittens could have been his.

Do cats love each other? They certainly seem to. The wives of the black-and-white tomcat lightly lifted their faces to his when they met him and seemed to want to be near him, especially when at rest. On cool mornings, he would choose the roof of the pickup, or a sheltered corner where the white barn reflected the sun, and, wrapped in his tail, would lie on neatly folded legs, his ears and whiskers relaxed, his eyes almost shut. Around him at various distances, perhaps on the hood of the pickup, one or two of his wives would lie in similar positions, not touching him, not even looking at him, but oriented toward him nevertheless and very much in his presence.

Not so our current cats, who today live in the same place—Rajah the deerstalker; Lilac, his mother; and Christmas, his mother's half sister. If the pattern above had no special meaning, one might expect that all cats would mindlessly copy it, and that Lilac and Christmas would orient to Rajah, to sleep where he sleeps, neatly curled and far apart but oriented to him nevertheless. But they do nothing of the kind. Although he is much larger than they, he is their junior, their son and nephew, not their husband. With spits and blows, they chase him away from their favorite places, and they respond with anger if he tries to chase them.

So I have come to believe that the male/female bond has much importance to cats. Female cats have their own ideas of what makes a good husband, and they exercise these ideas when choosing and supporting their mates. After all, the stronger the tomcat, the more successful he will be in defending his kittens, so that the mother's efforts to bring them into the world will not be wasted. When a cat family is settled and in place, with no males lurking on the periphery to threaten its safety, its members are relaxed and loving. In the population of barn cats that I knew best as an adult, the cats who belonged to my son's family in northern New Hampshire would usually greet each other, tails high, by fondly brushing noses when they met, a greeting initiated by the lower-ranking cat. The young cats grew up under a benevolent male despot, Rollo, who, although he was neutered and was not their father, functioned in place of their fathers. Their real fathers must have been several intact males from neighboring farms and households, but such was Rollo's presence that he kept these males at bay. Perhaps his lowered testosterone level smoothed things out, letting him refrain from fighting them when they appeared on the scene. The arrangement was excellent for the females, who got to raise their kittens without having to worry about marauding males. And, like many cat fathers, Rollo was excellent with the kittens, whom he groomed from time to time and allowed to play with his tail.

The impression that cat fathers mean much to their families was borne out by later studies of some British cats,* who did what the farm cats of New Hampshire had done so long ago. While the British male was confined in a crate for unexplained reasons, the females did not forsake him but spent the night ly-

---

* D. W. Macdonald, P. J. Apps, G. M. Carr, and G. Kerby, "Social Dynamics, Nursing Coalitions and Infanticide among Farm Cats, *Felis catus*," *Advances in Ethology* (Berlin: Paul Parey Scientific Publishers, 1987).

ing on top of the crate. Where he was, there they wanted to be also. Housecats similarly orient to their human owners or sometimes to the empty clothing of their human owners, showing a sense of oneness even if to the dog family and to the primate family it seems underdeveloped and vague.

Considering the affection that cats in groups show for one another, one might almost consider it the tragedy of the cat family that whether they want to or not, most young cats sooner or later move away from their group. Practical considerations drive them away. Except around farms, where the presence of stored grain keeps mice and rats in sufficient numbers and where the farmer supplements the food supply, the cat's diet doesn't encourage life in groups. Spacing is essential, and cats feel it. They don't need a hostile stepfather to make them move—sooner or later they'll be forced to move anyway. We were able to experience this with our own cats in a way that surprised me—they saw developments in our household that impelled emigration, while we, the human beings, saw nothing at all.

Our cats all came from that population that lived in and around our son's house and barns. These cats were all related to each other through the female lineages, with neighboring males arriving to father new litters every once in a while. Our first cat, the white male Orion, had been born to our son's high-ranking cat, the matriarch Manas, which may have enhanced Orion's status in the group. His high rank may have explained his unusual confidence. Even as a kitten he dared to climb to the barn rafters in pursuit of swallows—he was the only kitten in that population ever to climb so high.

When he came to live with me and my husband, he was our only cat, and he made our house his house, spending all his time with us except when he went out to hunt, which he did by means of a cat door. About a year later, we also adopted his half sister, the inky Wicca, who at the time was about four months old, a lanky adolescent. Orion was older than Wicca and much bigger, and we expected that Orion would dominate her. But to our surprise, she evicted him. Within a week or so of her appearance he had abandoned the house except at mealtimes and had claimed as his territory the building that houses my office, a barnlike structure that we also use as a garage, tool shed, wood-storage area, and home to several communities of mice. We might have gone on this way indefinitely, with Orion in the outbuilding and Wicca in the house, but we were given a third cat, the young Aasa, a cousin of the others although she had never met them. Aasa was the youngest and smallest of the three, but she took over the house in the same way that Wicca had done, and Wicca, by now a mature and very capable animal, moved into the outbuilding with Orion.

What motivated these emigrations? Not sex, since all our cats were neutered, which meant that reproduction played no role whatever in their territorial arrangements. Cat behavior is so strongly motivated by individual choice that we at first tended to shrug off their arrangements as personal preference—another cat mystery that human beings cannot solve. But then visitors came to stay with us, bringing their cat, and Aasa moved out. When the visiting cat left, Aasa moved back. Clearly, something was happening that the cats understood and that we did not.

At last it came to me what the cats might be doing, how the situation might appear to them. Possibly they saw me in the role of mother cat. If so, each new cat might have seemed like the kitten of a new litter, a signal for the older kittens of the previous litter to strike out on their own. What makes this phenomenon most interesting is that wild or feral kittens sometimes stay to share space with the mother. And when they do, they almost certainly have been invited by the mother. But what might constitute an invitation, I didn't know. So I didn't give it. So perhaps my cats, in a very typical barn cat manner, moved to the outbuildings to give me and my "new offspring" plenty of room. There they stayed for the rest of their lives, although we tried everything, including imprisonment, to keep them with us in the house. They'd stay as long as we prevented their moving, but the moment the chance came, they'd go, leaving our house to young Aasa. Barn cats from a world of fields, barns, sheds, and outbuildings—they did what they knew to be right.

On the island of Cyprus, in connection with a human habitation, archaeologists discovered the remains of a cat that dated from 7000 BP,* making it older than the earliest known domestic cats of Egypt. No wildcats lived on Cyprus even in those days—this cat had gone by boat. He was a big individual, wildcat sized. Was he merely tame, or had the domestication of cats gotten started? No one knows, of course. All that can be said for sure is that his relations with the boatmen were such that when they set sail for Cyprus, they brought him along.

How did we ever manage to domesticate cats in the first place, let alone so long ago? With their promiscuous ways and polygamous marriages, their intolerance of leaders and their exclusive living habits, cats are quite unlike our other animal slaves. Ironically, the only cat that could be said to meet, however superficially, the ordinary requirements for domestication is the lion. Yet before we rush to domesticate lions we might ask ourselves why most circuses that train big cats mostly train

* Juliet Clutton-Brock, *Cats Ancient and Modern* (Cambridge: Harvard University Press, 1993), p. 26.

tigers only. Why so few lions? The answer is that 700,000 years of group life have made lions highly assertive, the feline response to competition in close quarters. Lions are to the feisty and difficult bobcats what tigers are to the emotionally more restrained lynxes. Trainers prefer tigers over lions because tigers are more tractable. Furthermore, as trainers are quick to point out, lions tend to band together against you, not with you, which makes them extremely poor candidates for servitude.

Of course, there are exceptions. For example, I know a trainer whose lionesses came to her aid when a male tiger tried to attack her during a show. Possibly the lionesses helped out because the trainer was also a female and was to some extent a group member—certainly more so than the aggressive male tiger could ever have been. Or the lionesses may have seen the difficulty in the ring as a potential threat to themselves and merely wanted to nip trouble in the bud. Whatever the reason, though, the event would not have encouraged anyone seeking to domesticate a lion, because cooperative aggression is not a quality people want in their slaves.

So how did we manage to domesticate a species of this improbable family? The answer is that we didn't. Today, it is true, cats are fully domesticated, showing the usual characteristics of reduced size, reduced brain capacity, and an increased tolerance for crowding. But cats came by these qualities by accident, because the domestication of cats was an accident, a by-product of the domestication of grass.

Yes, grass. When we learned to harvest some of the grains, which of course are the seeds of several kinds of grasses, we accidentally harvested an entire little part of the ecosystem with them. Like grass seeds everywhere, the seeds of our ancestral grains would have provided natural food for a host of small species, among them the local species of murids, the rat and mouse family. Not surprisingly, when the grains were harvested, the local mice and rats followed their food indoors. And who should follow the mice and rats into the granaries but their natural enemy who had been hunting them all along, the local wildcat, *Felis sylvestris lybica*, soon to become *Felis catus*.

Domestic grains then spread throughout North Africa, Asia, and Europe. Along with the grain came the mice and rats. Behind them crept the cats, probably encouraged by the people, who by then had surely learned that cats are not only better mousers than dogs but also are cleaner and quieter and don't steal most kinds of human food. Certainly, cats didn't eat the wheat products that were becoming a main source of nourishment for the people and their livestock. The same could not be said for dogs.

In time, people invented ships and began to trade, using grain not only as a provision but also as an item of barter, and the mice and rats went along, once again followed by the cats. In this way, still linked after thousands of years, a food chain that began in the grasslands to the south and east of the Mediterranean Sea has spread to all corners of the world.

Today, like little islands, our homes preserve that food chain. Grain born in Asia Minor makes our bread. The house mice and the brown rats and black rats who feed from that grain live in our walls and basements, very different animals in behavior and in appearance from the wild mice and rats who live in the woods and fields. And the former wildcats sleep on our beds. Even the old names have followed the cats on their journey from their grassy plains so long ago—the word *puss* seems to be a version of Bast, the Egyptian goddess; the word *tabby* seems to come from the Turkish *utabi*, a striped cloth; and the word *cat* seems to come from the Arabic word for the animal, *quttah*.* Yet much has changed. Mice and rats have grown bigger and stronger over the centuries, and cats have grown smaller and weaker, which shows, I think, what natural selection has done for the rats, and what the softening, minimizing process of domestication has done to the cats.

A very informative study done at Johns Hopkins University on the feral housecats of Baltimore showed that the streetwise cats

* F. E. Zeuner, *A History of Domesticated Animals* (New York: Harper and Row, 1963), p. 390.

of Baltimore made a real distinction between rats who lived in the city parks eating grass seeds and other wild foods and rats who lived in the alleys eating garbage. The park rats were wild, so to speak, and as such received only average nourishment. Of normal size, they were perfect victims for the feral cats. The alley rats, on the other hand, were nourished beyond all expectation by the uncollected garbage provided by the people of Baltimore and were so massive and strong that the cats cringed away from them.*

The same was surely true of the rats on the New Hampshire farms when I was a girl. Like most rats that infest our habitat, ours were Norway rats—actually not originally from Norway but from eastern Asia. Grain for the livestock was stored in closed wooden bins which the rats opened with their teeth, gnawing holes in the rear where the people wouldn't see. Then the rats ate so well that they grew enormous, too big for the undernourished farm cats to tackle. Ferocious in defense of their well-ordered society, secure in their system of trails and passageways inside the barn walls, the rats seemed valiant and could be genuinely dangerous. Even the dogs were wary of them. In this context, humans often found the cats less than satisfactory.

On the other hand, the human beings on the farms weren't always entirely satisfactory to the cats. Farm cats, after all, are neither pets nor livestock. They have no monetary value whatever and are tolerated with mild amusement as long as they intrude only by occasional visits to the dairy at milking time to beg for milk. Barn dwellers, these cats were traditionally the province of the men, not of the women, and few male farmers had much commitment to them. At least, they didn't in rural New Hampshire, in the areas known to me in childhood. Barn cats were seldom fed by human beings; they were never neutered or spayed, never given any immunization shots, never taken to the veterinarian (who in those days in rural communi-

---

* James Childs, "And the Cat Shall Lie Down with the Rat," *Natural History*, June 1991, pp. 16–19.

ties specialized in livestock, and whose duty was to maximize a farmer's income, not to coddle his animals—the veterinarian of our community, for instance, pioneered the notion that anesthesia was a necessary part of surgery, to be included in the charge, and was no longer to be seen as an optional frill). When the cat population got too high for a farmer's liking, the cats were simply put into bags and gassed or drowned. To care for a group of animals for a time, and then to suddenly round them up and dispatch them without warning, is after all what farming is all about.

But if our relationship with cats has been only partly satisfactory to both sides, the relationship has also been less rigidly controlled than our relationship with our other domestic animals. All the others originally lived in highly structured groups with leaders. In animals with leaders, the leadership system is self-promoting in that good leadership benefits the group, resulting in a higher rate of survival than would be found in groups with incompetent leaders. Imagine a herd of wild goats living in the mountains under the leadership of a wise doe who knows about avalanches and who, on the spring migration to higher elevations, can lead her younger sisters, daughters, nieces, and grandchildren across steep slopes by knowing when to walk on the dangerously tilting snow. Saved from being swept off the mountain, her kindred will show a high rate of survival, and if the tendency to obey a leader is transmitted in the genes, the wise doe's kindred will pass the tendency down through the generations.

In order to domesticate such animals as the goat—as we did early on—all that we human beings needed to do was to insinuate ourselves into the position of their leaders. Our enslaved subjects, the animals, had been primed by their own excellent leaders to follow and obey.

But except for lions, members of the cat family don't follow leaders in the same way. Kittens follow their mothers, to be sure, but after that cats make their own plans and decisions, each individual taking full responsibility for himself or herself, with no one else to show the way. An adult wild cat may get some help from her mother in the form of a gift of land and even in the form of baby-sitting services and personal protection, but the normal day-to-day decisions a cat makes on her own. What, then, ties cats and people together?

In fact, cats and people are tied in several ways, the most important probably being the simplest—that of ownership. We own cats as we own all other property—they are ours to do with as we please. A few mild laws protect them from our tendencies to cruelty, yet these laws are easily evaded and are seldom enforced. Interestingly enough, however, as we own cats by human rules of ownership, so cats own us by cat rules of ownership. As a wild cat owns a territory and the rights to hunt the mice or the deer thereon, so a housecat owns a human dwelling and the rights to the people. People are not prey to housecats, it is true, but we provide food even more readily than would a mouse population. This is why cats mark our homes with spray and lightly rub their scent glands on our bodies. *Mine*, say the delicate odors, and other cats keep off. My husband's cat, Rajah, has staked out my husband's office where they both spend the day (unless my husband is away or Rajah is out hunting), and Rajah has claimed my husband's side of the bed, where they both sleep. Rajah has also staked out my husband, who by cat law belongs solely to him. One of our other cats, Christmas, who is Rajah's aunt, keeps far away from my husband. Even at suppertime, even in a snowstorm or a rainstorm, even at dusk when the cats come in for the evening, she won't come to him when he calls. To her, Rajah's territorial claims far exceed my husband's wishes in importance. Rajah's claims even exceed her own need for food and shelter. The other cat, Rajah's mother, also respects her grown son's property but sometimes overrides his wishes and visits my husband anyway. I, in turn, am shared by the two females, one of whom owns the house and the other the outbuilding that contains my office. They both share my person and take turns sleeping on my side of the bed. Rajah is very nice to me, very cordial when we meet, but since I evidently belong to his aunt and his mother, he maintains a respectful distance most of the time. He is more likely to visit me when my husband is in the same room, or when my husband is away altogether. Interestingly enough, the three cats share the land around the house among themselves, so that any one of them may hunt anywhere. Cats from other households very seldom intrude, but if they do, they are driven off by the dogs.

A second tie between people and cats is probably the most basic, that of parent and child. Many people treat cats like children, and therein must lie much of their cats' appeal. An ordinary housecat weighs six to nine pounds, is soft, warm, and yielding, and when held against the chest feels very much indeed like a human baby, or at least it does until it puts out its claws. But the tactile resemblance isn't the only resemblance. Like a human baby, a cat has a high voice, a small chin, large eyes, and a head of short hair that stands on end. Such powerful auditory and visual stimuli trigger atavistic care-giving behavior in our kind.

And if we treat cats as children, they treat us as parents. The length of an adult human being from the knees up is about three times that of the body of an adult cat, or, in other words, is in roughly the same proportion to the body of an adult cat as the adult cat's body is to a kitten. And people, too, are relatively soft and yielding. No wonder cats orient to people who are lying on their sides or sitting in chairs with their feet up. The curve of the human body must remind a cat of old times, of comfort and security in the warm curve of its mother's body with siblings all around. Many if not most cats enhance the effect by pedaling or kneading with their front paws, a well-known phenomenon sometimes known as "making bread" that kittens use to stimulate the flow of milk in their mothers' breasts and which they accompany by purring.

Even being fed by a person must seem like old times to a cat, because of the person's manner of delivering food. A person characteristically puts down a dish of food and moves away from it, offering plenty of space, which invites the cat to approach and eat. In the same way, a hunting mother cat puts down the dead bird she has brought, backing away from it to show that she will not compete for the carcass and that her kitten can approach. This system of food delivery is totally unlike the wolf-style dog system, whereby the providing adult lowers his or her head and heaves up a stomachful of chopped food while being mobbed by the puppies who eagerly lick the provider's mouth. Thus, the feeding method of the wild parent explains why dogs are ready to gobble when their bowls of food hit the floor, while cats tend to hang back and approach slowly.

Yet the paradigm of a family in which the human being is the parent and the cat is the child is only half of the story. As cats see it, they, too, play the role of parent while their owners play the role of child. Many cat owners relate how their cats have saved their lives, accounts which I, for one, was at first reluctant to believe. Some of the accounts hold that the cats wakened their sleeping owners because the house was on fire or for other environmental emergencies. These accounts seemed reasonable, but the altruistic motives ascribed to the cats seemed unconvincing. If the house was on fire, why wouldn't a cat wake its owner? How else could the cat get out?

Then I heard an account that seemed impossible to discredit—that of the scientist Lisa Rappaport, who not only is a biologist but who had been hired by an important zoo to make observations of animal behavior. It seems to me that if her account cannot be trusted, none can. One night in her apartment, her cat prevented her from entering a room in which an intruder was lurking. Lisa tried to go around the cat but the cat persistently blocked the way. Lisa would have had to physically remove the cat to get by her, which she wisely decided not to do. That cat had clearly assumed the role of parent/protector, not the role of child.

Perhaps in the role of parent or educator, cats may sometimes try to teach us. Our Lilac tried to discipline my husband when he shifted his position on a sofa, accidentally squeezing her. Calling out sharply to get his attention, Lilac cocked back a front paw, and when my husband turned in surprise to look down at her, she hit him. Interestingly enough, her claws were out, but not far out. She meant to prick him, not to tear his skin.

He was wearing a heavy sweater and didn't even feel the blow but was impressed nonetheless and has been more careful since.

Our son's cat, Rollo, tried to punish our granddaughter Zoë, age five, whose habit was to stroke Rollo's tail when she passed him, asleep, on his shelf above the radiator. Evidently Rollo didn't like this, but he tolerated it anyway, until one day he decided he'd had his fill. Down from the shelf he sprang, right on top of Zoë, and began to slash at her with both front paws, pummeling her like a boxer. She screamed and her parents came flying, chased off the cat and picked her up, but found not a mark on her entire body—Rollo had sheathed his claws.

Cats may be assuming the role of educator when they bring prey indoors to their human owners. Perhaps they want to feed us. Perhaps they want to teach us. Dead prey suggests feeding, but live prey suggests teaching in a number of ways. A mother cat starts teaching her kittens from the moment they start following her. They first learn from watching, crowding right up beside their mother if she lets them (a behavior that can lead to cooperative hunting), and later she gives them hands-on practice by flipping victims in their direction, exactly as a cat does in play. Mother cats even bring live prey back to their nests or dens so that their homebound kittens can practice, especially if the prey is of manageable size. So perhaps cats who release living prey in our houses are trying to give us some practice, to hone our hunting skills. I am speaking, of course, of the endless series of birds, mice, bats, voles, shrews, moles, snakes, frogs, toads, salamanders, hornworms, dragonflies, and other broken, twitching victims that our cats bring inside and set free.

Meanwhile, the practice itself raises some interesting questions. If, for instance, cats really see us as kittens, how old are we supposed to be? Not tiny kittens, surely, as these are fed with milk. Cats rarely lie on their sides and offer their breasts to human beings. No—we must seem like adolescents unready for life on our own. And if so, no wonder. Certainly we must seem unskilled if not downright clumsy, and we must seem impolite as well, since we show cats few of the courtesies that they show to us and to each other.

When our male cat, Rajah, had already grown much bigger than his mother, Lilac, he hadn't yet seen that as a reason to give up his perfect childhood. So at night he would curl himself comfortably beside my husband on the bed while his mother went out hunting. For our benefit he would purr loudly, but he would also pause from time to time and wait, ears up, listening for his mother's return. At last, in the darkest hour of the night, from the farthest corner of the house, he'd hear the quiet click of the cat door closing. Very alert, head high, he'd wait, still listening. Soon his mother would call him, with a booming, echoing *meow*. Before the echoes died, he would have launched himself off the bed with a powerful backward kick against my husband's body and would bound down the stairs into the dark kitchen where he'd find his mother beside the little corpse. Then she'd step back and he'd crouch down and polish off his food. Sometimes he'd even growl, evidently at her, the only living creature near him. But, in keeping with her indulgent mothering, she never objected.

It was at this time in her life and in this context that she brought food to my husband and me. She also brought food to our adolescent dog, Pearl, who had grown up with Lilac in our son's household. Three times Lilac fed young Pearl, each time placing a dead mouse in Pearl's bowl. Twice the bowl was on the floor, but once it was on the kitchen counter waiting to be filled at mealtime. Perhaps Lilac found the experience rewarding. Unlike Rajah, Pearl didn't growl at her, and unlike me or my husband, who tend to shout in disgust when provisioned by cats—we throw their gifts in the garbage—the dog ate Lilac's delicacies gratefully.

PART TWO

The Old Way

A nimals have no culture," wrote Konrad Lorenz. If an ethologist of his stature believed that animals have no culture, how many other people must think so too? But on the contrary, animals most certainly do have culture. We fail to realize this for no better reason than that our experience with populations of wild animals is so severely limited we are not often in a position to see much evidence of culture. Worse yet, we have been conditioned to believe that if we have seen one group of elephants, say, or lions, we have seen them all, so we don't even search among wild populations for cultural differences. But in fact, many if not most kinds of animals have culture just as we do, often very easily observed.

What is culture? Not what a professor of English critical of my claims once suggested. "Animals don't write poetry," he insisted. "They don't play musical instruments or attend the ballet." True, and they don't extend their little finger when they hold a cup of tea. But the concept of culture that I have in mind is more like what an anthropologist might offer—a web of socially transmitted behaviors. Dictionaries usually assign this definition to human beings only, but dictionaries are not the voice of God and will change their definitions eventually, as evidence unfolds.

Culture in animals comes about in precisely the same way that it comes about in human beings, by each generation learning from the generation before. A cultural solution to a certain problem gives the cultural participants the benefit of their ancestors' experience. Thus, culture makes life a little easier. Because everybody doesn't have to solve every new problem for himself or herself, but instead can look to his or her elders and colleagues as role models or for advice, culture functions much like instinct, benefitting its participants by smoothing the way. Naturally, cultural solutions to the problems of any given species, whether human or nonhuman, vary considerably from place to place. This variety creates cultural enclaves, and these make up the cultural units that we human beings perceive among ourselves. What we don't perceive so readily is that other species have cultural units too.

Many scientific studies tease away at the question—a landmark study in Great Britain, for instance, showed that two populations of chaffinches from two different areas sang two different songs. And chaffinches from one area who were raised in the presence of chaffinches from the other area sang the song appropriate to their foster home, rather than the song that their close relatives would be singing far away.

Culture in social animals is relatively easy to understand and to observe—wolves in different areas, for instance, show marked preferences for different kinds of prey. In some areas, the wolves are deer specialists, and in others, moose or caribou specialists. In still other areas, the wolves specialize in the difficult task of hunting beavers, even though other prey animals abound.

Culture in cats, supposedly solitary animals, is less obvious, yet it is there just the same. I witnessed an excellent example in my own cats, which came about quite by accident but could not have been more apt an illustration of culture if it had occurred by design. As I have noted, all our cats come from the same population, the population that lives in and around our son's home.

Over time, all but one or two of our son's cats were spayed, but not surprisingly, from time to time the unspayed cats had kittens. Some were kept by the family, and some were placed in good homes elsewhere. But now and then the family would feel that it had a cat too many, and would ask us to take the extra one. These cats were so healthy, so loving, so capable and intelligent, that we seldom refused. They may well have had different fathers, but they all came from the same lineage of females and thus are not radically different genetically, or not different enough to explain, in my opinion, divergences in their behavior.

As I have mentioned earlier, in the past whenever a new cat arrived from our son's household, the cat who happened to reside in our house made way for the newcomer by moving to an outbuilding, which was unnecessary in our household, but would have been very sensible behavior on a farm. Or at least this was what happened as long as our son and his family lived in New Hampshire.

But after a number of years, our son moved his family to Boulder, Colorado. On moving day, to the surprise of some of his neighbors (customarily farm cats are simply abandoned), he and his family took all the cats—loading them into carrying cases and boxes and putting them into a trailer which the family pulled behind the car (having carefully vented the exhaust with a hose, so as not to asphixiate the animals). All arrived in excellent condition, and all adjusted well to Colorado. There were still occasional kittens, and from time to time we adopted one. Our gray cat, Christmas, came to us from Boulder as a young subadult, and at our house met her relative, Aasa. In keeping with the old farmyard tradition, Aasa seemed to feel displaced, and moved to the periphery.

But there the practice stopped. More cats have been added to our family, but never again did a new immigrant cause an older cat to move away. Why not? Probably because Aasa was the last cat to come from the old barnyard culture. Although all the new cats came from the same population as the old cats, that population was by then solving its problems in a very different setting. No longer did their property include barns and out-buildings. Rather, all the cats born to that particular population after its transfer to Colorado grew up in and around a suburban house, with no more land than a back yard, and no outlying place to take shelter without venturing beyond the group's territory. As a result, when the newer litters of kittens reached adolescence, if they weren't given to new homes, they simply stayed on. And when one of them found himself or herself in a new setting, such as our house, he or she kept the customs learned from kittenhood, from observing the behavior of the older cats. A minor matter, perhaps, yet it wouldn't seem minor to the cat family, since one of the most significant acts in a cat's life is to leave its natal setting and find a new home.

Nowhere is cultural difference more evident than in pumas, who in certain areas have learned to leave livestock alone.* The Idaho pumas I was privileged to observe, for instance, spent their days in the presence of unsupervised livestock on an open range. From their high lookouts the pumas could actually watch not only the adult horses and cattle but also their very tempting, easily captured foals and calves. Even so, simple as it would have been for these pumas to help themselves to some of the livestock, they passed them by and instead hunted deer. Moreover, a rancher whose livestock grazed safely within the sphere of such abstinent pumas would enjoy their protection, since in defending their own territories the pumas would keep possible killers of livestock away.

Supposedly, this should have contributed to the survival of abstinent pumas, since the ranchers, in theory anyway, should have realized that they weren't losing livestock to pumas, and appreciated what was going on. But American ranchers are notoriously indifferent to and ignorant of these matters, so that Idaho and Utah have lengthy hunting seasons for pumas, and the game wardens make only feeble, ineffective efforts to inhibit poaching. Hence the pumas lose their lives to hunters anyway.

* Personal communication from Ken Jafek.

How does one observe cultural change in a cat population? Probably the best way is to revisit a population over a period of time. I was privileged to revisit the lions on the Namibian side of the Kalahari Desert, first in the 1950s and later in the 1980s. Originally I was a participant in an ongoing study my family made through the courtesy of the Ju/wa Bushmen who live there. Although the focus of the study was to learn about the Bushmen, my own interest was in the lions. I had little time to indulge that interest in those busy days, but lions are hard to ignore, so I was able to learn something about them. But the big revelation came when I returned to Nyae Nyae in the mid-eighties, only to find that while the people had changed greatly, the lions had changed even more, so that many of the things that I thought I had learned about them no longer held true. Also, the old lion population has been divided. Where there was one lion population, now there are two, each as different from the other as they are from the population I had known.

All this had happened in just thirty years, or, from a lion's point of view, in less than two lion lifetimes. Plenty of lions still living could have been reared and educated by lions who had known the old days, and there had not been enough time, by any stretch of the imagination, for the lions to have undergone genetic adaptation.

In the 1950s, the lions and the hunter-gatherers were the most formidable beings in the Kalahari. This fact was widely recognized. "Where the lions and the Bushmen ask for your pass" was a synonym for the Kalahari in apartheid times, when passes were required of black South Africans and when non-Bushman people seldom ventured beyond the limits of the farms and the towns. In those days, the Kalahari Desert was a vast wilderness, unexplored by anyone except the Bushmen. No safarists, no travelers, no farmers, no white or Bantu people had ever stayed in the roughly ten thousand square miles of dry bush savannah that formed the most remote parts of the western desert. The Bushmen were the only people ever known to have lived there. Few non-Bushmen had even passed through it. Most of it had not been mapped. There were no boot marks in the dust, no roads across the plains, no jet trails in the sky, no satellites among the stars. In those days the western Kalahari was still whole, a delicate ecosystem of enormous antiquity. The plant communities, untouched by modern man and his domestic animals, were drought adapted and in balance. Their size and composition were controlled by time, by soil conditions, by rainfall and fire. The populations of animals, in keeping with a dry savannah, were not large and crowded, as they are in most African game parks today, but sparse and highly varied. Showing the ancient relationship between this place and some of its inhabitants, the antelopes who lived there had become independent of water and could range throughout the great dry areas of the western Kalahari, getting enough moisture from dew and from little wild melons called tsama melons. In those days, water-dependent ungulates such as rhinos, elephants, zebras, and buffalos were not present in the western Kalahari except as occasional visitors during the rains. But all the large southern African carnivores were generously and widely represented— an indication of the health and stability of the ecosystem.

The only primates in the western Kalahari were the people, the Ju/wasi. When we visited these people, they were living in the old way, as hunters and gatherers, getting their food, their clothing, their tools and their shelter from the savannah. Tobacco, small pieces of metal, and tiny glass beads were the only exotic materials they had, and they got those from the neighboring Bantu-speaking peoples by trading animal skins. The Ju/wasi smoked the tobacco in pipes made of animal thigh-bones, and they cold-hammered the metal into knives and arrowheads, gradually replacing their old style implements and bone arrowheads with the new material, though the implements remained unchanged in usage and design. Otherwise, like the Kalahari itself, Ju/wa technology was stable. The most recent innovations seemed to be the bow and arrow and an extremely powerful arrow poison made from the grubs of *Diamphidia* beetles and their parasites.

No one knows how long the Ju/wasi or any other people

have lived in the western Kalahari. No one knows whether the modern Bushmen are descendants of an ancient population or are newcomers in a series of occupying populations. Needless to say, in the 1950s no archaeological work had been done on them. Excavations in bordering areas have since unearthed encampments containing artifacts of hunter-gatherers—encampments that were occupied more than thirty thousand years ago. That figure was determined while the excavations were in their early stages. At the time of this writing, the depths of the encampments have not been reached nor the antiquity of their occupancy established.

Although these archaeological findings will certainly be important to our understanding of human prehistory, they may be unimportant ecologically. What mattered to the integrity of the environment was that human hunter-gatherers had been there long enough to count as ecologically indigenous. Human beings, after all, evolved on the African savannahs. Fossil evidence from sites such as Olduvai Gorge places some of the earliest hominids just a few weeks' hike to the north. So people and their ancestors could have been gathering the groundnuts and wild melons of the western Kalahari for a very long time.

During that time, the human populations were governed by the same forces that governed the populations of other living things. The ecosystem absorbed the impact of our species just as it absorbed the impact of, say, the lions. As a result, the hunter-gatherer economy as it was practiced by the people in the Kalahari will probably prove to have been among the most successful human economies ever practiced on earth, if duration and stability are any measure of success.

Even the plants of the Kalahari seem to have adjusted to our kind. Although plants have been adjusting to animals since the time of the dinosaurs and before, their adjustment to fire may have been furthered by people, at least on the African savannahs. In the Kalahari only people and lightning can set fires, and lightning sets only a few. The rest, the fires whose smoke used to redden the sunsets at the end of the dry season, were of human manufacture, set from time immemorial by hunters to induce the new green grass, which draws the game.

To some Westerners, the practice of casually setting a fire and then letting it burn on to travel where it will throughout the dry season until the rains come and put it out might seem destructive. Yet in fact, at this point in its evolutionary history, the Kalahari's vegetation profits from the fires. Without fires, a certain kind of thornbush takes over. With fires, the grass is renewed and many plants germinate. The fires help the vegetation, which in turn helps the animals.

Although the people of the Kalahari got more of their food from gathering than from hunting, and thus were very different from the other carnivores, the animals saw the people as hunters and acted accordingly. As the antelopes responded to lions by signaling warnings or by positioning themselves to keep the lions in view, they responded to people by staying beyond bowshot—a distance most antelopes seemed to know. In fact, some animals seemed to know a lot more about arrows than just their range; many giraffes knew to keep the branches of a bushy tree like a shield between their chests and a person. That defense would do nothing to help against any other predator or against any other weapon, but against the small, lightweight arrows used by the Bushmen it worked very well. Lacking the right kind of tree, a giraffe would move farther away from a hunter.

So much knowledge and such highly developed safety strategies on the part of the game meant that the human hunters and the other hunting species faced the same difficulties: the game was almost but not quite a match for them; they were pushed to the limits of their skills. In terms of hunting success, it seemed to me, the Ju/wa men were probably more or less equal to the other large-sized hunters of the Kalahari—especially those who hunted large-sized game cooperatively. In those days, the most important of the other large-sized cooperative hunters in the western Kalahari were, of course, the lions.

Whatever large animals the people hunted, the lions also

hunted. Unlike the many predators who have no choice but to kill old or young or sick or wounded animals, the Bushman hunters simply killed the nearest animals. So did the lions. The people tended to favor giraffes and the larger antelopes—hartebeests, ghemsboks, kudus, and elands. So did the lions. Even the hunting technique of the people resembled that of the lions. In contrast to other large-sized coursing hunters such as cheetahs and members of the dog family who have the stamina to run down their prey, the people for the most part crept near their victim through some kind of cover and then sprang out for the kill, just as cats do. A lion springs from hiding onto his victim's back and, twisting its head around, clamps shut its windpipe, stifling it, while a person springs from hiding and throws something, or shoots a poisoned arrow, but although the actual kill is achieved differently, the underlying lurk-and-spring method is the same.

Perhaps these shared preferences aren't surprising; both people and lions needed the same kind of victim for the same reason—to feed a group. Most of the time, the people preferred to stay in groups of twenty or thirty, while the lions apparently preferred to stay in groups of six or seven. Hence, in a way, the group size was the same: the people were more numerous, but the lions were larger. Each group would have weighed about three thousand pounds, so on a cosmic set of scales the groups would have balanced each other. A meat meal big enough for the people was also big enough for the lions.

Perhaps partly for this reason, the lions and the Ju/wasi helped themselves to each other's kills. Once, I was present when Ju/wa hunters robbed some lions. As we were walking along in the bush, the hunters noticed vultures coming down out of the sky. When we went to the place where they were dropping, we found white-backed vultures in a tree above a rack of red bones that once had been a hartebeest. Since white-backed vultures are not birds who delay gratification, their sitting in the trees meant that lions were very near. Realizing this, the hunters approached slowly. The lions must have been watching from the bushes, but they didn't object. The hunters picked up the carcass bit by bit, deliberately and confidently, if mindful of the lions. No lion showed herself to challenge them.

Another time, my brother, John, was present when lions tried to rob people. He and four Ju/wa hunters had been following a wildebeest that one of the hunters, days before, had shot with a poisoned arrow. When, at last, they caught up to the wildebeest, it was lying down on folded legs in a clearing in heavy brush, very ill from the poison and surrounded by an unusually large pride of lions and lionesses—about thirty of them. Some were subadult, but many were mature lionesses in their prime. Back in the bushes was at least one mature maned lion. The four Ju/wasi took in the situation, then slowly advanced on the lionesses. Speaking firmly but respectfully, they announced that the meat belonged to people. The lionesses rumbled unpleasantly. Some stood their ground. But others turned tail and retreated to the bushes. And then, although the bushes seemed alive with huge, tan forms pacing and rumbling, the Ju/wa hunters descended on the wildebeest, tossing clumps of earth at the lionesses, speaking firmly and respectfully as they did so. At last, the lionesses slowly, unwillingly, backed off. As soon as the lions and lionesses were screened by the bushes, the Ju/wa hunters seemed to give them little further thought and turned their attention to the wildebeest; they surrounded it, killed it, skinned it, and cut it into strips to carry home, leaving nothing behind but a green cud of partly chewed grass.

These days, this story must seem incredible, so it is fortunate that the event was well documented. My brother, a filmmaker, had a loaded camera with him and filmed everything. Yet at the time, I saw nothing remarkable about a group of four Ju/wa hunters and my brother, armed only with clumps of sod, chasing thirty lions from their intended prey. I was taking my cues from the Ju/wasi, to whom the encounter seemed almost a matter of course. Naturally, I was deeply impressed by the courage of the Ju/wa hunters—only another Ju/wa hunter could take that for granted—but I should have been equally impressed by the lions.

I thought we were finding out about lion nature. I thought

that this was how lions would behave anywhere if approached as the Bushmen approached them, with firm but respectful requests, without fear. After all, the Bushmen had seen nothing unusual about the event and had known what to do immediately. Nor were the lions unusual, for that time and place. We all knew, of course, about lions who might have acted differently—for instance, the lions who lived at the edges of the Kalahari where Bantu people kept cattle. But at the cattle posts, because the lions hunted cattle, the people hunted the lions. One of the Ju/wa hunters explained the different behaviors by saying, "Lions are dangerous only at the cattle posts. The lions around here don't harm people. Where lions aren't hunted, they aren't dangerous. As for us, we live in peace with them."

That was certainly true, but it was not the whole story. Rather, it was the Ju/wa side of the story, which in those days was the only side ever given any consideration by any human being. All of us assumed that the people, not the lions, determined the events.

But the lions also had a share in shaping the relationship. A truce if ever there was one, the people-lion relationship wouldn't have worked unless both sides had participated. Yet how and when the truce started, and what the lions gained from it—and therefore what they put into it—have never been precisely determined.

The beginnings would have been very deep in the past. Our ancestors evolved on the African savannahs in the presence of lions, who themselves had been in southern Africa for 700,000 years. Perhaps we were too small to interest them as dietary items. We were most certainly too small to take any kind of commanding tone with them in the manner of the four Bushman hunters' attitude toward the thirty lions as witnessed by my brother, and we were too small to chase them away if they wanted our food. If there was robbery in those days, the lions, not we, were the perpetrators. Obviously, by the 1950s, something had changed.

It is often assumed that such a change in the relationship between people and animals is due to the development of

weapons, but the Bushmen's weaponry wasn't necessarily superior to the lions' in a combat situation. The Bushmen had their
spears and their bows and arrows, marvelously designed for
hunting, yet all but useless in self-defense. The Bushmen's
spears were then, and still are, lightweight and barely four feet
long—or about the length of a lion's reach—far shorter than the
formidable nine-foot heavyweight spears that the East African
pastoralists used when testing their manhood against lions.
Bushman bows and arrows are also very small and light. To do
its work, a Bushman arrow does not need to pierce a vital, in-
ner organ but needs only to pierce the skin, to inject a drop of
poison anywhere at all inside the body. But a drop of poison
could take several days to kill a lion. Meanwhile, the hunter
might find himself in considerable danger if not torn to shreds.
In short, Bushman arrows are about the worst possible weapons
for self-defense.

So if not arrows, what? No one in his right mind would
think of going up against a lion with a knife. Nor would fire be
a useful weapon. The Kalahari animals are accustomed to fire;
they evolved in the presence of fire, and they are not afraid of
it. Anyway, the normal bushfire in the Kalahari has little to burn
except grass, and isn't very hot. If caught by a fire, animals and
people alike merely wait for the fire to come to a patch of low
grass and then step over the flames.

Also significant is the fact that, unlike the East African pastoralists, the Bushmen have no shields of any kind, and never
have had, as far as anyone knows. A shield, after all, suggests
that the owner is expecting trouble. The warfaring, lion-hunt-
ing Maasai, for instance, carried enormous shields. The Bushmen, in contrast, seemed to expect no trouble. Skinning knives,
lightweight bows and arrows, short, lightweight spears—hunting tools all—were the only weapons the Bushmen had or felt
they needed.

A better explanation for the truce—the only good one I can
think of—is that the people, who were not combative with each
other, were also not combative with animals. People hunted, of
course, but hunting isn't a form of combat—or at least it wasn't

to the hunter-gatherers. Hunting was merely a method of obtaining food and clothing. Most animals, as a rule, avoid conflict when they can because conflicts cause injuries, and injuries impair survival. For most of our time on earth, our kind, too, had to abide by the practical considerations that govern other animals. And the Bushmen in the 1950s lived in the old way, by the old rules.

Yet the observation of a truce did not imply that the Bushmen took lions lightly. On the contrary, the Bushmen thought of lions as they thought of no other animal. The Bushman hunters whom we knew deeply respected the hunting ability of lions and even said that lions are better hunters than people because they understand teamwork, in contrast to the Bushmen, who hunt singly and only work as a team* when tracking an animal after it has been shot with a poisoned arrow. Teamwork, said the Bushmen, allows the lions to exploit the tendency of hoofed animals to circle back for a look when followed by a hunter. With a team, one hunter could position herself (if she were a lion) or himself (if he were a person) to be waiting where the prey animal turns back.

Also, the Bushmen recognized a supernatural quality in lions. The word for lion—*n!i*—like the name of God, could not be uttered in the daytime. And people attributed human qualities to lions. Certain lions, for instance, were thought to be sorcerers, a belief that may have come to the Bushmen from the Bechuanaland Protectorate (Botswana, after 1966), where certain people were said to take the form of lions by night, but to be unaware of the transformation. When lions formed a line at night and kept in touch with each other by sequential roaring, the Bushmen sometimes said that the roars were emitted by a single supernatural lion sorcerer, a werelion, who took great leaps through the air, giving a roar each time he touched the

---

* There is no word for *team* in !Kung—the phrase used was *//ue //om*, "do it together."

ground. A lion could even cause an eclipse of the moon, said the folklore. He would cover the moon's face with his paw to give himself darkness for better hunting.

The success of the Bushman/lion truce seemed truly remarkable. While we were in the Kalahari in the 1950s, we knew of only one Ju/wa person who had been injured by a lion—a man who had been mauled while helping a group of Herero ranchers hunt a cattle killer in the Bechuanaland Protectorate. Being a Bushman and a servant, he had been conscripted as a foot soldier in the Herero-lion wars and during the advance on the lion had been forced to the front with the dogs. There the lion had mauled him. We knew of no one who had been killed by lions. This fact became all the more impressive in light of a genealogical study made in the early 1980s by my brother and a colleague, Claire Ritchie. The study, which attempted, among other things, to determine some of the causes of death among Ju/wa Bushmen, took in more than three thousand people and went back about a hundred years. Naturally, people remembered deaths by unusual or violent causes very well, and among such deaths they recalled several caused by animals, mostly snakes and leopards. But among fifteen hundred deaths recalled by the hundreds of people whose testimonies I was able to examine, only one was said to have been caused by a lion. The victim in this account was a paraplegic—a young Ju/wa girl. My daughter, also a paraplegic, is a source of intense interest to captive lions and tigers and has only to roll her wheelchair through a zoo to get the big cats bounding in their cages. She proposes that the girl's motion—slow, uneven, and low to the ground—caused lions to regard her differently from the way they regard able-bodied people.

Furthermore, throughout the world attacks on human beings by the cat family are most often made on people who are low to the ground—people squatting to relieve themselves are well-known targets, as are children. The sight of a head peeking up above cover such as grass has a mesmerizing effect on all cats, as our cat Rajah demonstrated recently while I watched a televised educational program about the sea. On the screen, a seal's head broke the surface of the water. Rajah, who had been watching the screen very closely, as if he knew that something was about to happen, flung himself at the head the moment it appeared, although he had already learned from many sad experiences with his cat video that televised images are not real. But he evidently preferred to take a chance on being wrong than on missing so golden an opportunity.

A head above a surface has equal appeal to lions, as a certain film company learned to its sorrow. Preparing for a shot of the underside and vanishing hindparts of a lion as it leaped over the cameras and ran for the horizon, the film company had dug a trench as a workspace for members of the camera crew, whose heads, moving back and forth just above the ground, gave the temporarily restrained lion something compelling to watch and to think about while waiting for his cue. When the lion was released, instead of jumping over the trench, he (not surprisingly) jumped at one of the heads and mauled the person badly. Fortunately, the person survived and evidently is still working as a photographer.

Even people with very inconspicuous disabilities are quickly zeroed in on by cats: once at a tiger act attended by some two hundred people tightly packed in a north-facing amphitheater, the entering tigers stopped in their tracks to stare into the sun at someone they had spotted deep in the crowd. Following their gaze I finally found what they had noticed immediately: a child with Down's syndrome sitting quietly and (to me) inconspicuously amid his family.

Just as the Nyae Nyae lions respected the lives of the Ju/wasi, so the Ju/wasi respected the lives of the lions. Although poisoned arrows are not the ideal weapon against lions, the Ju/wasi would certainly have used them without hesitation if it had become necessary to rid themselves of a problem lion. Keeping out of the lion's reach until it collapsed would have been inconvenient but by no means impossible. No—only the fact that the lions gave the Bushmen no cause to want to harm them explains the forbearance. The lions kept their side of the truce.

In contrast with the lions were the other members of the

genus *Panthera*, the leopards, with whom the Ju/wasi seemed to have a somewhat different relationship. I attended one leopard robbery, during which the Ju/wa hunters simply collected a leopard's kill without showing him any respect, not even a cautious approach—they just picked up the carcass and left without so much as a backward glance. But the leopard had kept himself discreetly hidden during the robbery and therefore hadn't required respect.

Leopards seem to have a somewhat ambivalent relationship to people. Shy creatures who tend to live in social isolation except for the usual mother-child families, they lack the boldness of lions even in areas where no one molests them. Even so, they are formidable animals, and they know they are frightening.

The fact that the Ju/wasi grouped leopards with hyenas and other predators who merited no special treatment was in itself interesting, because it suggested that the purpose of the respect shown to lions was not necessarily to mollify them. On the contrary, if the Ju/wasi had wanted to mollify an animal, the leopard would have been the logical choice, for among all the large animals of the Kalahari, leopards were the most dangerous to people. While we were there, we heard of several people who had been killed by leopards, and the survey made by Claire Ritchie and my brother revealed several more.

Leopards are smaller than lions and, unlike many lions, live alone or in very small groups; therefore, leopards are more apt to be satisfied with small and mid-sized prey. Size of prey is of real importance to the cat family, even to housecats, particularly as the object of a hunt. A housecat, say, will catch a beetle and may even chew it up and swallow it, but only as a snack; the same cat would probably not start his hunt with a beetle in mind as his quarry but rather would be envisioning something larger. Why waste energy hunting something too small?

A sad fact for human beings is that our size makes us almost perfect prey for leopards, a characteristic which the leopards have not ignored. Leopards haunt the camps of human beings, and this habit has led to their scavenging people's temporarily unburied dead. Sometimes, during an epidemic of terrible illness such as smallpox, all the people of a group would be stricken at the same time. No one would have the strength to bury the dead, and thus the bodies would have been available to scavengers. According to Ju/wa eyewitnesses, the practice of scavenging seemed to lead leopards into watching camps in which many people were ill and, in a few instances, into taking the very ill shortly before death. Possibly this habit also led some leopards into entering camps in which no one was ill— something they may have done in the past and may still do today, killing people and, if possible, carrying them off. In the most recent episode I know of, which was in 1987, the leopard didn't even wait to choose someone who was asleep but took a man who was sitting by a fire. His wife beat off the leopard and saved him. We didn't hear of anyone's being taken by a leopard in any other way than from a camp at night; no one was stalked or ambushed while gathering roots or berries in the bushes, or while crouching down to urinate in long grass (a practice which in the interests of privacy is used by Ju/wa men as well as women), or while crouching down to get water in the tall, thick reeds that surround some of the waterholes. Leopards every-where hunt other animals by those methods, and some leopards in other areas hunt people by those methods. So the restriction of man-hunting to campsites at night was apparently a cultural feature of the western-Kalahari leopards—leopards whose tra-ditions went back to epidemics of the past.

Why didn't the lions do likewise? Didn't they, too, have the opportunity to learn of a possible food resource in the camps of sick or sleeping people? Hungry lions should by rights refuse nothing, and in many places man-eating by lions is well known. On certain roadsides in East Africa, for instance, lions occa-sionally wait for drunken people to come staggering home at night from bars. Because man-eating is so often practiced by single lions, by old or young male lions who evidently lack the advantages of pride membership, it is my opinion that man-eating is a lion's solution to the problem of landlessness. A lion without land must poach on the territory of other lions. If the nomadic, landless lion captures one of us, the resident lions won't mind as much as if he had captured an animal such as a kudu or a ghemsbok that they, too, could most likely have eaten some day. Thus a nomad's wish to avoid competition, to avoid attracting the negative attention of other lions, could make human beings seem like tempting prey.

Not in the Kalahari of the 1950s, though. Not in that deep and perfect wilderness. No one can explain the truce because no one understands it. The truce was simply taken for granted, as most situations involving animals are simply taken for granted. Animals are assumed to be static in nature. So even today, with both the human and the animal populations stressed and damaged, few people realize the difference between how things are now and how things were then.

During the years that we stayed in the Kalahari, we often lived near a waterhole called /Gautscha. One of three permanent waterholes in the area, it lay in a rocky outcropping in a thicket of long reeds. For much of the year, /Gautscha was the only source of water in nine hundred square miles of very dry country.

On a rise of ground a few hundred feet east of the waterhole, in the shade of a grove of little trees, the group of Ju/wa Bushmen camped in shelters made of bent branches sparsely thatched with grass. We camped nearby in tents. To the west of the waterhole lay a great clay pan, /Gautscha Pan, which formed the bottom of a shallow lake during the rains but was a bare, cracked mud flat in the dry season. In a dry bank to the northwest of the pan, perhaps a mile or two from the waterhole, were the dens of some spotted hyenas. Somewhere nearby, until someone killed him, lived a brown hyena. On a neighboring grassland lived a cheetah. In the vast bush southwest of the pan lived a leopard. And in the (to us) featureless bush to the southeast lived a pride of lions. Its size varied, but there were never fewer than ten. We never found the lions' resting places, nor did we try to find them, but we thought we knew where they were because we sometimes heard lions there in the morning, when lions tend to gather together, and in the evening, when, after a day's rest, lions begin to move around. So in an area of a few square miles lived about thirty people, ten or more lions, a cheetah, a leopard, and at least five hyenas, or approximately fifty large, predatory creatures, all of them hunting the same antelope population, all of them drinking from the same waterhole.

Helping to minimize the chance of meeting was the habit of the different groups to use the area and its resources at different times—the people and the cheetah by day and the other predators by night. Time of day was particularly important for the people and the lions because the people needed daylight for hunting and also for gathering, and the lions, who couldn't hope to hunt if they couldn't conceal their large bodies, preferred darkness. The grass was seldom long enough or thick enough to hide them by day.

As one group spread out to forage, the other group would gather together to sleep. Further limiting the chance of meeting was that neither group started the day's or night's activity quickly. The lions began their hunting not at dusk, when the people might still be on their way home, but long after dark; the people, on the other hand, delayed leaving their camp until the day was well along and thus never met the lions—or, for that matter, any nocturnal predator who might be finishing a night's hunt in the dawn.

Yet, for all the factors that kept the groups apart, we often did meet the other predators. For instance, we often heard or saw the hyenas. Watching at night by the waterhole, I would see them when they came for a drink. Unlike the hyenas in game parks, these were not used to vehicles, and, eyeing my jeep with great suspicion, they would stalk around it like cattle who have seen a dog. But they were not shy about visiting us. One night, while poking stealthily around our camp, one of them very cautiously put her head into my little backpacker's tent. I was reading with a flashlight and looked up to see her sensitive nose just inches from mine. Our eyes met. "What is it?" I asked. Unsure, she drew back. We would also see hyenas when we went by their dens, which had been dug, like caves, into a vertical bank. One hyena, a large female with breasts, would stand half in, half out of her doorway, watching us with an unfriendly, almost twisted facial expression, as if she found us repugnant.

Occasionally, we would see the leopard stretched out upon a certain rock, his thighs loosened, sunning his furry white loins. We would often see the head and shoulders of the cheetah above the grass when we went to his flat grassland. But we never, in all the miles we traveled on foot or in vehicles, on all the nights we spent watching for nocturnal wildlife by the waterhole, chanced upon one of the lions.

I used to ask people what would happen if someone met a lion in the bush. If that should happen, I would be told, one should walk purposefully away at an oblique angle without exciting the lion or stimulating a chase. Several times, people showed us how to do this. But at /Gautscha we never met a lion. Although among us we spent at least parts of more than fifty person-years in the bush there, we never once had occasion to use the technique we had learned.

We saw it used, though. One day, in the close quarters of some heavy bush in the farthest waterless reaches of the Kalahari, my brother and I met a lion. He was all golden in the sunlight, with a golden mane. He seemed very large and, unlike many Kalahari animals today, he was in beautiful physical condition: he had no scars or scratches and had plenty of flesh on his bones. Stupefied, we gazed at him, in awe of his presence and his beauty. He stood still, gazing at us. How long we might have stayed this way I don't know. My brother and I were too dazzled to do anything. So the responsibility fell upon the lion. Moving calmly, confidently, purposefully, keeping us in view without staring at us aggressively, *he* walked purposefully away at an oblique angle. The effect of the encounter on us—or at least on me—was memorable. The lion was only a few feet away, and I could have become afraid for my life. Yet his intentions were so clear and his demeanor was so reassuring that I felt absolutely no fear, not even alarm—just interest and wonder. By his smooth departure and his cool, detached behavior, the lion apparently intended to save himself the risks of an unwanted skirmish. A man acting in a similar way under similar circumstances would have been considered refined, gentlemanly, polite. In our species, too, reassuring manners can bring desirable results, for exactly the same reasons.

That was the only time, as far as I can remember, that any of us saw a lion by chance. But it was not the only time we saw a lion. In fact, we often saw them. That, however, happened at their discretion, when they wanted to see us. Usually this happened on the first or second night we spent in a new area, or upon returning to a familiar area that we hadn't visited for a long time. As animals who keep close watch over all that happens in their sphere, the lions would inevitably come to check us out.

These days, my own three cats remind me of the Nyae Nyae lions whenever an unfamiliar vehicle stops in our driveway. At first, the cats scatter, perhaps to await developments or to consider the situation while remaining out of sight. After a while, they emerge inconspicuously from hiding, then discreetly advance upon the vehicle. Before long they are sniffing its tires, and when they have satisfied themselves about the odors thereon, they jump to the hood or the roof and simply stand there for a time, neck long, ears up, tail down, hindquarters low, hind legs bent at the knee as if waiting to see who will come to challenge them. They do this in all seasons, day or night. It seems to me that the cats are intrigued by the height, and wish to experience the vehicle as a high thing, the way that one of us might climb on a pile of lumber or a mound of hay. Later, if the door is open, the cats go inside and explore the interior. It is as if they are asking themselves what difference the vehicle is making to their world. We try always to ask visitors to close the doors of their vehicles when no one is in attendance, lest a cat get shut inside and driven away. (This has never happened to us, but it has happened to our neighbors. They were lucky— the owner of the van, an electrician, realized whose cat he was transporting and very considerately drove him home again.)

Our first encounter with lions was on our first trip, at the edge of the Kalahari, far to the west of /Gautscha, just after we had camped for the night. With us was a young Afrikaner man, a former smallpox-control officer, who had come to show us the way to a place where he once had found and vaccinated some Bushmen. (In those days, almost no non-Bushmen had contact with the Bushmen, or even had any idea where they were.) In the dark, a group of five lions came quietly up to us. Beyond our fire we saw their shining eyes, which were so high above the ground that we thought at first we were seeing donkeys. When I realized that we were seeing lions, I was overcome with excitement and ran around the fire to see them better. Just then, a bullet whizzed by my ear, shots rang out, and the eyes vanished. Before anyone realized what the young Afrikaner was doing or could stop him, he had shot two of the lions.

That was all he did, too. He wouldn't even go to see if he had killed them. When the rest of us found tracks and splashed blood but no dead lions, we realized the extent of the problem created by the young man—two wounded lions nearby in the dark. We asked him what he was going to do about it. Nothing, he said. It was, after all, nighttime. It would be dangerous to follow up the lions. So the task fell to me, my brother, and a man named William Cam, who had come with us as a mechanic.

We set off on foot in the starlight, moving very quietly so that we might hear the lions breathing or the low, mumbling growl that a wounded lion might make. We also tried to catch their scent. At last we heard a soft moan. We followed the sound, turned on the flashlight, and found a lion—a male, full grown but still too young to have a mane or to have left the pride. Badly wounded, he was lying on his side, unable to get up. He was evidently in pain, for he had been biting the grass. We had to shoot him several times before we could kill him and each time a bullet hit him he cried. One of the worst moments of my life, that scene is as fresh in my mind today as it ever was, and as painful. The lion turned his head aside, to look away from us as we stood over him and shot him. I wonder now if by averting his gaze he was hoping to limit our aggression.

We couldn't find the other lion, and after many hours of searching we gave up, to try again in the morning. When the sky grew pale, at the place where the lions had been when the young man fired, my brother and I found the tracks of a lion who had taken a great leap. Not fifty feet from camp, at the end of the next leap, lay the body of a lioness shot through the heart. She, like the lion, seemed young: she still had spots on her white belly. Her fur and the grass around her were cold and wet with dew. Or mostly cold and wet with dew. Right beside her we found a warm, dry place where the grass lay flat. Looking around, we saw a dark trail through the grass where something had knocked off the dew. Then on the trail we saw a grass stem starting to rise after being pressed down, then another, and another, and under the slowly lifting grass stems we found the round footprints of an enormous lion, who had left only moments before. So we knew that while the dew fell, this huge lion or lioness had stayed beside the dead lioness, within sight of our camp, listening to all our comings and goings, listening to the shots and cries. During the night the watching lion or lioness had groomed the body of the dead lioness, turning her fur the wrong way.

Our next encounter took place on the first night of our second trip to /Gautscha. We had come in vehicles after much hard traveling. We were too tired to pitch tents so, about fifty feet from the Ju/wa encampment, we threw down our sleeping bags and without even bothering to build a fire went immediately to sleep. During the night, we heard the Ju/wasi saying some strong words to someone, but we didn't pay much attention. We were too tired. In the morning, we found the footprints of lions all around us. Several lions had come to investigate us as we slept and had even bent down to sniff our faces. We found the huge, round prints of lions' forefeet, toes pointed at us, right by our heads.

Afterward, the lions had gone on to the Ju/wa encampment and had stared over the tops of the little grass shelters at the people there. Unlike us, who stayed awake all day and slept all night, the Ju/wasi took naps during the day and got up often at night. Hence they were virtually never all asleep at the same time. Even in the depth of night someone would be awake, getting warm by a fire, having a snack or a sip of water or a chat with someone else. When the people who were awake saw the burning green eyes, they got smoothly to their feet and firmly told the lions to leave. Since the Ju/wasi would hardly take a low, commanding tone with one another, the unusual voices woke everyone else. At first, the lions didn't want to leave, but the people insisted, and at last shook burning branches at them. Eventually, the lions went.

On several occasions, lions seemed to have strong feelings about us, about something we had done or were doing. As I look back, the interesting thing about the episodes is not that they were frightening, which they were, or dangerous, which they could have been, but that the lions seemed to be trying hard to communicate with us, perhaps simply to give expression to their feelings, perhaps to make us do something.

Unlike the lions, who correctly understood, and even obeyed, the spoken and gestured commands of the Ju/wasi— words and gestures that were designed for other human beings and then merely applied to lions—we human beings were not able to understand the lions. Not even the Ju/wasi understood them, and they knew them better than anyone else. Why could the lions of /Gautscha understand the requests of the people but the people not understand the requests of the lions? Are lions better than people at understanding interspecific messages? Are people better than lions at conveying messages? No one really knows. It came to me, however, that our kind may be able to bully other species not because we are good at communication but because we aren't. When we ask things of animals, they often understand us. When they ask things of us, we're often baffled. Hence animals frequently oblige us, but we seldom oblige them. Elephants are different, but then, elephants can motivate people as no other animals can. Once, an elephant who didn't want me near him threw gravel at me so hard it felt like buckshot. I understood at once what he wished to communicate, and thereafter I paid scrupulous attention to his bound-ary, which was, incidentally, not the bars of his cage or the edge of the sidewalk but a creation of his own mind and seemed to be expressed by an unchanging but invisible line.

Of course, each time an animal tries to communicate with a human being, the animal is pioneering, since there are no es-tablished ways. Sometimes the animal tries something that is familiar and that works with his or her own kind—a dog who wanted something might, for instance, bark or stare or whine, all ways in which he might successfully communicate with an-other dog. But not all animals are satisfied with the familiar. Cats are particularly inventive in communicating with human beings, and most of us can see plenty of examples in the efforts of our own cats. Our cat Orion, perhaps having noticed that at night I would investigate noises on the stairs, once jumped hard from step to step, and repeated the procedure so successfully that I thought the sound was being made by a heavy person, and I got out of bed to investigate. Orion had, I saw, been jumping on the top three stairs only, and when he got to the third step from the top, he would go back up and do his jumps over again. He was just starting over for the fourth or fifth time when I ar-rived. He then looked at me and *meowed*. He had food, he had water, he had a cat box. In short, he lacked nothing that I ordi-narily would provide for him. It occurred to me that he wanted to go out, and needed me to open the cat door for him. But we had long since begun keeping our cats indoors at night, for fear of a coyote, and I couldn't oblige him. I said aloud, "I'm sorry, but I can't let you out." His eyes lingered on mine, as if he were taking in what I had told him, and then he turned his head and went on down the stairs in perfect silence. Did we communi-cate what I believe we communicated? Had he really wanted to go out? Had he understood my remark, or some of it, and de-duced the rest from the tone? Possibly—he was certainly com-municating something, and I may or may not have picked it up.

But in my experience, the most dramatic episode involving a cat's attempt at communication took place one hot, moonless night in the rainy season at /Gautscha, when a lioness came to our camp. Most of our people and also many of the Ju/wasi were elsewhere. I happened to be alone in a tent in our camp, and my mother and brother happened to be visiting people in the Ju/wa camp, about thirty yards away. I was working on my notes by lantern light. At the Ju/wa camp, about six small fires burned. We had been in residence there for almost a year and in no sense could be considered newcomers.

At about ten o'clock that night a lioness suddenly appeared between the two camps and began to roar. The loudness of lions cannot be described or imagined but must be experienced. My body was so filled with the sound that I couldn't think or breathe, and in the brief silences between roars my ears rang. The earth and the walls of the tent seemed to be shaking. Terror-stricken and confused, I tried to collect my wits. There was nowhere to go that gave more protection than the places we were already in—I in a very flimsy tent but at least not completely exposed, the other people all together beside fires. Climbing a tree was out of the question—there were no trees whose upper branches the lioness couldn't reach by standing on her hind legs. At last, with trembling hands, I carried the lantern outside the tent, partly so that its light would shine on the tent rather than through the tent, to make the fabric seem solid rather than transparent, with me quivering inside like a shadow puppet. I also wanted to illuminate the lioness so the other people could see where she was, because her roars were so deep and so loud that they gave no direction. To judge from her roars, she was all around all of us—anywhere, everywhere.

She seemed to have in mind something in the Ju/wa camp, since she was looking in that direction. She seemed not to notice the lantern. With her ears half up and turned sideways, with her tail taking great, full sweeps, she seemed angry and edgy, a lioness whose patience was at an end. Sometimes she would pace back and forth, and once she leaped out of the lantern light, only to leap back into it again. It is sometimes claimed that lions roar at other creatures to confuse or stampede them, making them easy prey. That night, such an explanation seemed improbable. Long ago, natural selection would have removed from the general population any people unwise enough to leave their fires and weapons and scatter in the dark, especially at the urging of a lion. Even Western people don't necessarily stampede under such conditions. Not knowing what to do or where to go, they simply remain rooted to the spot with terror. That was what happened to me. As for the Ju/wasi, the lioness certainly got their attention but perhaps didn't frighten them as badly as she frightened me. Cool but alert, they awaited developments. Anyway, there wasn't anything anyone could do. The lioness certainly didn't seem in the mood to consider a firmly spoken request from the Ju/wasi, and that night they didn't offer any; they maintained a tactful silence. It seemed to me terribly important to notice how long the lioness stayed there, so I timed her. She roared intermittently for almost thirty-five minutes. Then she left, with swift, impatient strides. And there the episode ended. She never came back, or not in any obvious manner, and no one ever knew what it was she had wanted of us.

Another time, lions combined their investigation of newcomers with unexplained roaring. The event took place on the second night my mother and I, along with some of the Ju/wasi, spent camped at Tsho//ana, a Kavango cattle post by a ravine about fifty miles north of /Gautscha. Up the ravine and into our camp came a great group of lions, and they began to roar in unison. Some began to roar as others ended, so that no gaps appeared in the appalling sound. Again the earth shook and the tent rattled. Noise so loud literally robs the breath and stuns the senses. We were paralyzed. At last, as suddenly as the noise began, it stopped. Then came a long silence, more terrifying than the roaring. The lions must have been listening, surely to learn the effects of their aggressive bellows. I held my breath and tried to keep my jaws apart so the lions wouldn't hear my teeth chattering. The lions apparently heard nothing and began to roar again.

As frightened as I was, I couldn't help pointing my flashlight's quavering beam around in the hope of sighting some of the lions. But they were right behind the tent, where I couldn't see them. Instead, out of the night, out of the deafening, thundering din, came one of the Ju/wa men. He had been on the far side of the ravine when the roaring began and, armed only with his little spear, he had crossed the ravine to be with his wife and children. Walking silently on bare feet, he had actually woven his way among the roaring lions in the dark.

W hat was the importance to the lions of the truce between them and the people, and what did the lions have to do to keep the truce? To consider the question, one must consider lion life. Probably the ideal situation for a lion is to live in a pride that owns a territory. Next best, perhaps, is to live with other lions in a nomadic, landless band. Least desirable, it may be, is to live alone as a nomad, although that is often done by young adults and by displaced males. But since a solitary life is possible, since a lion can hunt for himself or herself and, if alone, doesn't need to share food (to judge from the way lions fight when they eat, they seem to hate sharing), why do lions prefer groups?

Surely one reason is that, like housecats, they care about each other. Mealtime behavior notwithstanding, like housecats lions rub faces lovingly when they meet, sleep near each other, groom each other, and keep in touch by voice when far apart. There are also many practical and economic advantages to group life. George Schaller, in his 1972 study *The Serengetti Lion*, reports that under certain circumstances lionesses hunting cooperatively in groups of five or six were more than twice as successful as lionesses hunting singly. And no wonder. A team of lions can hunt more effectively than a single lion, and while the carcass must then be shared by all the hunters, so that no lion gets nearly as much at a sitting as she would get if she hunted alone, she nevertheless gets to eat more often. All in all, by providing moderate meals at reasonably regular intervals as opposed to enormous meals at rare intervals, teamwork is the better way.

Nor are lions the only cats to hunt in groups. In fact, other cats do, too, probably for many of the same reasons. Housecats sometimes hunt together, with at least rudimentary cooperation if not with the disciplined teamwork of lions. As I have mentioned, my cats even hunt cooperatively with my dogs, or at least they tolerate the presence of the dogs, since the cats usually initiate the hunts and the dogs join in when they see what's happening. (Such hunts are often quite successful, since two animals limit the avenues of escape, thus entrapping the little quarry. Usually the cat makes the catch but then relinquishes the corpse to the dog, who eats it, just as a lioness might relinquish her victim to a lion.)

Like housecats, lionesses cooperate in child care. It is not unusual to see a lioness striding along with cubs of different ages stumbling behind her. Lionesses nurse one another's cubs as cats do and even teach one another's cubs. My brother once came upon a lioness holding a struggling warthog in her front paws while two large cubs and two small cubs looked on, very interested. From the bushes, a second lioness watched. In short, as long as food is available, there are few disadvantages to group life for lionesses and virtually none for lion cubs. Even if a mother is away for long periods, she can know that her infants are safe, held close between the warm thighs of their aunt or their grandmother, fed by her four black nipples, and protected by her brave lion heart.

Some lionesses live alone as nomads, but do so probably because group life is not available to them. Such a lioness was Elsa, in *Born Free*. When George and Joy Adamson, who reared Elsa, tried to get a pride of wild lions to accept her, no pride would take her—any more than a family of hardworking people would take in a homeless stranger simply because the stranger was of the same species. The saddest thing about Elsa was that as far as she was concerned the Adamsons had been her group and, from her point of view, they forsook her.

A male lion also derives many advantages from a group. Although he begins life in his mother's group, he leaves when he reaches young adulthood as all cats do, probably driven out by a new stepfather. The young lion's task, then, is to find a group that will accept him, a pride in which the lionesses will not be his mother, aunts, and sisters but his wives. Finding such a group is not easy, if for no other reason than that most groups already have a resident male or males. Until the young lion can find a group whose resident male is missing or is too old or too disabled to defend his lionesses, the young lion lives as a nomad. Yet occasionally some find vulnerable groups, defeat the resident male, and take over. In return for the sexual and hunting services of the lionesses, the lion guards their territory, if they have one, and fathers their children until other young males appear and defeat him. Then, if he survives the battle, he must once again become a nomad, squeezing out an existence at the fringes of other lions' territories until he dies. It is easy to see what great advantages a male lion can derive from a companion—another male to join his battles, to share his responsibilities as well as his wives. Normally, young males—sometimes littermates but sometimes just friends—find themselves pushed out of their group or groups around the same time. Joining forces, they find new lionesses. Together they defeat the resident male or males, together they defend a territory and father the cubs. In fact, a group of males provides a much more stable situation for the rearing of cubs than a single male can hope to do, since the group members can better defend their position. Not easily can newcomers take over from them.

With all the advantages to be had from group life, it is surprising that some lions live singly even where game is plentiful. Solitude seems to be the choice of some lions in the desert reaches of the southern Kalahari and in the sand dunes of the Skeleton Coast. Could the determining factor be water?

Water is seldom considered as a possible factor in the size of lion groups, for in many parts of Africa lions are known not to drink water. The moisture they need evidently comes from the bodies of the animals they kill and also from dew and from tsama melons. Even so, I believe that water as well as the alternative forms of moisture may be more important to lions than we now suppose. Lions are not like the antelopes, whose special physical mechanisms enable them to tolerate great heat and to limit their bodies' water loss. Lions are like us—they must cool themselves. Just as we lose moisture when we sweat, lions lose moisture when they pant, and they must replace it. For a possible answer, we may look once again to the Kalahari hunter-gatherers, whose group size at certain times of year may have been related to the availability of water. During the dry season, in places where there was no water, where the people got moisture from plants and from the bodies of animals, the groups were small. Where there was water, the groups were much larger. Almost seventy Ju/wasi had communal rights to the waterhole of /Gautscha, and although the food supply didn't encourage so many to live there during the dry season, more than that number would meet there for periods during the rains.

In contrast were certain people who, perhaps more than any other people on earth, were attuned to desert life: the /Gwi Bushmen. Before we went to /Gautscha, we visited the group of /Gwi Bushmen mentioned earlier, the people who were living without water. They had no hidden spring, no secret sip wells or buried ostrich-eggshell water containers—not even a pool of stale rainwater in a hollow tree. Like desert lions, this particular group of two men, four women, three teen-age boys, and one baby got liquid from a number of different kinds of watery plants and from the bodies of animals. From the latter these people were past masters at collecting every drop. On one occasion, when we were able to measure the amount of liquid, we found that they collected about five gallons from the rumen of an adult female gemsbok, one of the large antelopes of the Kalahari. To do it, they made a bowl of the gemsbok's skin, so that no liquid escaped. This technique gave each person several good drinks. Many more people could have eaten from this gemsbok, but not many more could have satisfied their thirst.

In the same way, the availability of water may have affected the size of the lion group at /Gautscha. With only the liquid in melons, meat, and dew, the lions might have had to scatter or live in small groups. So the waterhole may have enabled them to live together. And they seemed to want to be together, as their pride of thirty shows. On the day that my brother and the Ju/wa men drove the thirty lions from a wounded wildebeest, the lions were far too many to have met by chance. That day they were together by design, and not as an aggregate but as a community. The appearance of a staggering, weakening wildebeest had surely focused them, but if they had been widely scattered, the Ju/wa hunt would have happened too suddenly to draw them from afar. They had to have been near one another when the wildebeest came along. And they weren't together for hunting; in fact, they were about twenty-five too many for efficient hunting. They had been brought together by some other impulse—a lion's impulse—and the water may have let them indulge it.

In the western Kalahari, since there were only a few sources of permanent water in the dry season, the vast savannah was not as vast as it seemed. Whoever wanted to drink water had to live within reasonable distance of a waterhole. In the case of the /Gautscha lions, that was the /Gautscha waterhole, and probably no other. We believed that there were lions associated with the other dependably permanent waterholes, which were many miles distant from /Gautscha. Other prides had surely established lion ownership of those waterholes. So it seems possible that the lions at /Gautscha, knowing that the source of water was unique, maintained a low profile around it. To mismanage their public relations so as to endanger their access to water could have altered their social opportunities. The lions contributed to the low profile by using the waterhole very late at night. They also came and went quietly. They didn't roar near it. They never lay near it, viewing it all day, as the lions in some game parks now do. And no lion ever left a scat by the water.

What do scats mean to lions? That might depend partly on their cat nature and partly on their particular culture. To me, the discretion of the /Gautscha lions about leaving scats near the water was reminiscent of certain housecats who won't use litter boxes that are placed too near their feeding areas. Perhaps these cats don't want to mix food or water with feces. Or perhaps scats mean more than simple waste to these cats. After all, most cats use scats as hostile signs. And scats seem to carry a negative message for lions as well. A resting lion, for instance, gets up and leaves the vicinity of its sleeping comrades if it wants to move its bowels. But under most circumstances, most lions seem to view scats with indifference. In places frequented by lions, including the waterhole areas of game parks, their scats are many and obvious. But the /Gautscha lions were different. Perhaps they wanted to safeguard the cleanliness of the water. Perhaps they wanted to avoid leaving negative messages. Yet whatever their reason, they viewed the waterhole area as an unsuitable place for feces and never soiled it. The lions knew why. We didn't.

The people, too, used the waterhole with care. They didn't pollute it or sit around it but drew water and left, usually at about the same time of day and never at night. And, just as the lions owned the water from the point of view of other lions, the people owned the water from our point of view. Other people wishing to use it needed permission. Thus, with respect to water use, the people and the lions had much in common, although neither group, perhaps, would have seen it that way.

In the 1950s, the lions of /Gautscha belonged to one continuous population, a single lion nation occupying a more or less undivided country. At the eastern edge of this lion nation were the cattle posts of Herero, Kavango, and Tswana ranchers. A Kavango family maintained such a post at Tsho//ana, where a lioness had worked out an amazing technique to secure cattle.

In response to the threat posed by the lions, the Kavango family had built a fortress of heavy poles, each about six inches in diameter and about twelve feet long, all planted in a deep circular trench that held two or three feet of each pole below ground and left nine or ten feet above. It would be hard to devise a more substantial, safer kraal. Yet in 1955 at least one lioness removed cattle from this kraal. How she did it was not exactly understood. Of course, her method didn't interest the Kavango family as much as her banditry did. She was seen only once, by torchlight, at the top of the fence, scrambling out with a heifer over her shoulder.

Was there only one lioness who could leave the kraal carrying a heifer? Were there several? Since only one lioness did it at a time, or on any given night, and since no human beings really knew much about the pride or its membership, no human beings knew or ever will know what actually happened. I feel safe in assuming that the lioness preselected her victim. She couldn't easily have made her choice while flying through the air, so almost certainly she studied the herd ahead of time, probably through the cracks between the poles. She would choose a mid-sized animal, perhaps a yearling heifer—not a large cow, which would be too big to carry, and not a little calf, which would be so small as to make the risk and the effort wasteful. Then this formidable lioness would leap from the ground to the top of the fence, wedge her forearms between the points of the uneven, tapered poles, and brace herself with her hind feet, her claws dug into the wood. From there she would probably locate her chosen victim. A quick scrabble would put her over the top and into the crowd of cattle, who, of course, would raise an appalling alarm, bringing out the Kavango men with weapons.

In among the dangerous sharp hooves and horns of the frantic cattle, the lioness would seize her heifer and, leaping to the top of the fence, would scrabble out again, bearing the entire weight of the heifer with only her shoulders and her mouth. It is hard to appreciate such a mouth: one must hold a lion's skull, taking time to admire the huge, arched, buttressed, deep-rooted eyeteeth and the wide, bony anchors for the massive jaw muscles. Two men together, using their entire bodies, could just barely carry what one Tsho//ana lioness picked up with her mouth.

A somewhat different situation pertained at the village of /Kai /Kai, a Herero cattle post about fifty miles east of /Gautscha. /Kai /Kai had been a Herero settlement for a long time, how long I don't know. If at first it was a place of Bushmen, a rich Herero found and took it, using it as a place to keep part of his vast herd. He seldom if ever came to /Kai /Kai. Instead, a few poor Herero families with their Ju/wa servants tended his cattle for him. The kraal at /Kai /Kai wasn't as strong as the kraal at Tsho//ana, although the people kept more cattle. Yet at /Kai /Kai, perhaps because the people were merely the guardians and herders, not the owners of the cattle, they put up less resistance to the lions, who had long been coming in groups to help themselves, until they had grown so bold that they would walk through the village in broad daylight while the people shut themselves in their mud and wattle houses. Only the village dogs—starved, beaten skeletons almost too weak to walk—regularly put up any resistance to the lions. When the lions came the dogs would rush at them, barking, and sometimes would actually succeed in chasing them off. Sometimes the lions wouldn't leave, though, but instead would chase the dogs. The dogs well knew the lions could kill them, so they tried to stay out of reach.

For the most part the people did little to stop the lions, but how could they do more? /Kai /Kai was in the Bechuanaland Protectorate, and in preindependence times the protectorate government forbade black people to have rifles or bullets. At /Kai /Kai the Herero men had to defend themselves with a muzzle-loading musket, for which they couldn't always get powder. But one day they had powder, so the gun was ready, loaded with nails, when, according to a Ju/wa participant, the lions strolled into town.

Armed with the gun, the people came out to confront the lions. The dogs barked as usual, but this time, encouraged by the people behind them, they broke with custom and rushed right up to the lions. Surprised at this change in the rules, the lions suddenly charged the dogs. The dogs turned tail and ran headlong for the protection of the people. Between the people's legs they scurried, with the lions right behind them. Most people were forced to scatter. But the man with the musket stood his ground and fired. The blast and the pattern of nails turned the lions' charge. The lions left and, as far as I heard, never came back in the daytime. No person was hurt and no lion was disabled that anyone knew of. Only a dog had failed to escape, a dog whose back had been broken by a heavy blow from a lion. A few people had noticed her, a paraplegic in a seated position, slowly dragging herself by her front legs. She didn't last long— she certainly wasn't there when I visited later. A few people remembered her predicament with amusement, but as to what became of her, no one knew or would have cared, except the other dogs.

In the 1950s, the lions around Tsho//ana and other cattle posts in the area had large, prospering territories that happened to include cattle pens that they visited only because they chose to. As time went by, however, the peripheral areas of what had once been the lion nation were slowly taken over by ranches and farms. In the Bechuanaland Protectorate, some years of drought tended to concentrate the huge herds of cattle in the few places where grazing could still be found. In other areas, farmers expanded their holdings or changed their style of farming, giving up the old-fashioned, unmechanized methods for more efficient, mechanized, commercially oriented methods. Gone were the days of a few sheep sent out to graze in the care of a dog and a child; in place of such inefficiency, large herds of livestock were guided through cycles of rotational grazing by men in vehicles. The end product, which had once been milk, a little cheese, and a little mutton for the farm's dependents, changed to commercially raised beef and hides for the export markets.

As a result, the peripheral areas of the lion nation became more precarious. The lions who lived there became the unfortunates of the population—the poor. The antelopes were quickly hunted out by the ranchers and farmers, for sport, for meat, for disease control, and to eliminate competition for the grazing. As the grassland was denuded and the once delicately balanced savannah became, in places, a moonscape with dunes, even the peripheral areas contracted. Then the lions had no other place to go and nothing else to eat: they were forced to hunt livestock. Of course, the farmers demanded the eradication of all lions. And the lions seemed to understand. Apparently in an effort to save themselves, the lions who lived on the periphery stopped roaring: the roaring alerted the people, who then hunted them more intensively. How did the lions know not to roar?

When I was in northern Uganda, where a similar situation

prevailed, the former lion range had in just two lion lifetimes been invaded by the powerful Dodoth, a Maasai-like people with big herds of cattle. As the cattle began to displace the game, existence became increasingly hard for the lions. They couldn't emigrate to better country because to the south, east, and west were crowded farmlands, and far away to the north, in a game reserve, where there was good, rolling country and abundant game, other lions already owned every inch of space.

Fugitives in their own home, the lions of Dodoth county tried to adjust to the presence of dangerous people by living as discreetly as possible—by leaving cattle alone except for strays, by staying out of sight, by moving only at night, by seldom even leaving their footprints on the roads, and by never roaring. Lions don't roar when they are trespassing on the territory of other lions, and for the same reason: the trespassers don't want the owners hunting them down. Therefore the Uganda lions acted toward the people as if the people were lions and as such were the rightful owners of the land, and as if they, the resident lions, were not the residents at all but were deposed lions or interlopers, although they were the only lions there.

To stop roaring must have been a real deprivation to the lions because the cat family in general and lions in particular do much of their communication by voice. To try to understand the probable cost to lions of not roaring, it seems useful to compare cats to dogs and to ourselves. The reason to communicate may differ with each species, but the benefit is the same. As a general rule, dogs come when called but cats answer. Dogs therefore follow the pattern set by their ancestors the wolves, who range very widely with the sun or the stars as their beacon and the pack, wherever it may wander, as their home. When called, a dog realizes that the caller may be afoot and moving and, rightly fearing permanent separation, hurries to join the caller, to catch up. An adult cat, on the other hand, is sure to be firmly ensconced on some kind of spot—his spot, however temporary—and he sees a call as a query, not as a summons. If he feels safe, he'll answer, which will disclose his whereabouts, but he probably won't feel the need to join whoever called. We, too, have atavistic feelings about calling and being called. We, too, feel the need to call to the members of our group when we're apart, and if we are called we feel impelled to come and also to holler "Here!" or "What?" Very few of us lack this response; it's a rare person who for no reason just stands there dumbly with his mouth shut. So when one thinks of one's own need to communicate and of how it feels to keep silent when one normally would call or answer, one can perhaps see what enforced silence cost the lions. Housecats, too, keep silent when they are frightened, and don't wish to be found.

As for the Uganda lions, their silence was worthwhile. To some extent, the low profile maintained by these lions helped them to survive. People didn't hunt them and hardly even thought about them. Not infrequently in northern Uganda, my own bad planning would force me to grope my way over long distances through the wildest country in the middle of moonless nights, usually alone, seldom with a flashlight, never with a rifle. Occasionally on these dicey journeys I worried about a certain leopard who haunted our camp and was evidently contemplating my children—age two and four—but I never worried about lions.

Once I almost met the leopard. It was a dark, overcast night, and as happened embarrassingly often I had gotten lost without a flashlight and had spent about an hour groping around in the dark bush until I finally recognized the silhouette of a hill against the sky. It seemed to be the hill we had camped upon, and as I was making my way up a trail that I hoped led to my camp, the leopard coughed loudly ahead. "Waugh!"

He was, I realized, on the trail. He had coughed, I reasoned, to reveal himself, to show his presence, to give me plenty of warning so that I wouldn't come on him suddenly in close quarters. I think his idea was that I should change direction and go around him. He certainly didn't want me walking up to him, getting surprised and all upset and perhaps involving him in some impulsive reaction at close quarters.

But I couldn't oblige him. I couldn't get off the trail, since without the trail I would gave gotten lost in the dark again and then would have been wandering vaguely in deep bush with a leopard close behind me. No—the alternative was to keep going, even though it meant making straight for him, which I did. For just a moment, he said absolutely nothing. Then, sounding startled, as if he couldn't believe I was simply going to ignore him, he coughed again, much more forcefully. "*WAUGH! WAUGH!*" He thought I hadn't heard him!

But my own situation remained the same—I was in the wild, dark bush very near to a leopard and I didn't dare leave the trail. I had to reach my camp and my children about two hundred feet beyond him, so I still kept going. The leopard fell silent to listen. He would have heard my steps falter (because I so badly wanted to do as he said), but then he would have heard me stride forward with renewed determination. So he said no more. Instead, still in absolute silence, he stepped off the trail and let me go by. How near he let me come to him I never learned.

The Ugandan lions, in contrast, kept out of the way of people so well I never had a close encounter. Occasionally at night when I would be driving to my camp in a Land Rover, I would come upon the entire pride sitting in the road. Like a flock of pigeons they would fly off in all directions to escape. It might have been interesting to see what, over time, the lions would do about their new situation, but most of the lions, the leopards, and the other large predators are gone, gunned down for sport, first by the King's African Rifles, again by the Uganda Rifles, and yet again by Idi Amin's army and the invading armies that came in his wake. I suspect that any lions who may be there now are recent immigrants from the Sudan.

etween the late sixties and the mid-seventies, more lions of Botswana and South-West Africa (now Namibia) were displaced, for the same reasons that the lions of Dodoth county in Uganda were displaced. In the developing commercial farmlands of South-West Africa, almost all the lions were hunted out. But not quite all; some lions were still able to cling to life on remote farms as lone, cattle-killing fugitives. One, a big male with a black mane and distinctive footprints, made a name for himself—Jakob—for his near-human powers of reasoning and his exceptional elusiveness. Over the years, all the white hunters and most of the ranchers and farmers shot at Jakob, but the bullets didn't hit. The failures seemed sinister to the would-be assassins: they didn't see his breathtaking escapes as possible indications of Jakob's education or intelligence, rather, seeking the explanation in their own minimal educations, they attributed Jakob's escapes to Satan, his alleged protector.

Sad to say, Jakob was killed in 1986, though not by any of the loud-voiced, beer-swilling hunters and farmers with their incredible lion stories, their high-powered rifles, their hollow-point bullets, and their telescopic sights. No, Jakob was killed by a Bushman with a thirty-six-inch bow and a quarter-ounce poisoned arrow—with just one shot. Only a Bushman hunter would know how to stalk a lion like Jakob in the moonlight—or have the nerve, for that matter.

No one knows what happened to most of the displaced lions. Perhaps some of them managed to invade the territories of lions in areas that were only lightly disturbed. If so, the general lion population might have declined even more. When a lion population is in a state of flux and movement, with territories being invaded and prides of lionesses being gained and lost, the infant mortality rises. And when the dust clears after the battling and the infanticide, any given territory left to lions holds roughly the same number of lions it has always held. That is true of any animals who must own land.

In what was the western remnant of the lion nation, by now separated from the eastern section by a belt of farms and settlements about a hundred miles wide, an effort was made to preserve some of South-West Africa's wildlife by the creation of Etosha National Park, which was formally established in 1958. To the eternal credit of the government, the need for a park had been recognized much earlier—mainly because of Etosha Pan, which the park encloses. A national landmark on the order of Old Faithful or Niagara Falls, Etosha Pan is an impressive, shimmering white salt flat in the dry season and a shallow lake during the rains and has for some years been an international attraction. By the early seventies, farms surrounded much of Etosha Park, which had become a wildlife preserve of about 8,600 square miles of mixed bush desert, grassy savannahs, and mopane forests with widely spaced, scrubby little trees.

Because the park was created on a map with a pencil, its existence as such wasn't noticed at first by its original occupants. They, of course, were the Bushmen and the game. While farms were laid out around the park, life within its boundaries went on as before.

But after the Second World War, the park authorities took notice of the Bushmen and reasoned that Etosha would be a better, more natural place without a human population. Nothing much was done for a time—in the early fifties, many Bushman families still made their homes near the waterholes of Etosha—but as tourism began to grow, as tourist facilities were built and roadbeds were scraped so the tourists could drive to view the animals, the Bushmen became undesirable. They begged from the tourists and they hunted the game. It wasn't nice to be sitting in the privacy of your car, enjoying the majestic elephants and getting the feel of the wilderness, only to see tapping at your window a half-naked Bushman smiling hopefully and pretending to puff the tip of his finger as he begged for a cigarette.

By the 1960s, almost all the Bushmen had been evicted but not necessarily resettled. Apparently, little attention was paid to where they went or what they did. Since they had no formal education, vocational or otherwise, since they spoke none of the European languages, and since they were generally unfamiliar with Western customs and with money, alcohol, machinery, buildings, vehicles, roads, cities, and farms, their futures were grim. Only a few Bushman men remained in the park as trackers and camp servants to the white personnel. A few Bushman families became laborers on neighboring farms. But all the other Bushmen, perhaps five hundred or so, seem to have disappeared. So ended the old way in Etosha.

By 1986, when I visited Etosha, the Bushmen had been gone so long that their former presence was beginning to seem romantic. At least one young white park official had begun to reconstruct the old hunter-gatherer past. Although some of the evicted people were surely still alive somewhere, the official ferociously enjoined visitors to the park from disturbing any of their old campsites because these, he said, could have archaeological value. The official had also written a paper in which he gave translations of the old place names. But not convincing translations. In his paper, a waterhole that might have had a simple, classic name such as Gu Na (Big Waterhole) became something like The Place from Which You Can See if Anyone Is Coming from Keitseb. This young park official had as his servant a Bushman in his mid-fifties, a man who was almost certainly one of the people who had lived in the old camps. Because he was about my age, and because when my family first visited in the 1950s we had traveled to the Etosha area, I wondered if I had ever met him. But I didn't get a chance to ask. The official did not favor interracial fraternizing. Wherever the official went in his pickup, he took along his servant, who rode in the back, as if in apartheid times, uncomfortable and possibly in danger, looking ragged and threadbare among the prosperous tourists and very much alone.

In the 1980s I went to Etosha for a reason that had nothing to do with the Bushmen, or even with lions. I went because a friend named Katharine Payne had made the important discovery that elephants make calls too low for people to hear—calls that travel great distances and by which elephant herds that are far apart can keep in touch. Before Katy's discovery, people jokingly spoke of "elephant ESP." Since her discovery, hardly a reference is made to elephants that doesn't mention their infrasonic calls. While Katy was working to prove the existence of these calls, she invited me to join her research team. The work, sponsored by the National Geographic Society and the National Science Foundation, brought us at last to the wild elephants in Etosha, where we spent two seasons.

At first, we worked at a waterhole in Etosha called Gobaub, beside which Katy and her research team built an observation tower. Gobaub, which was far from the areas that tourists were allowed to visit, reminded me of /Gautscha. A wide, flat plain like the pan at /Gautscha surrounded the waterhole. The plain seemed to be the bottom of an ancient lake. Around its edge the old lakeshore rose to a rolling sandy expanse of heavy bush and mopane forests. The waterhole itself was a wide pool formed by a spring bubbling out of a ledge. The pool and its runoff provided drinking places for many animals, and their long trails approached it from all directions, like the spokes of an enormous wheel.

One night, as Katy and I were leaving the observation tower to go back to camp, we heard a great ruckus of roaring and screaming and of pounding, running feet. I shone a flashlight beam and saw, at a distance, a herd of about ten wildebeests facing a group of four lionesses, three of them ranged in a line behind the fourth, for all the world like three backfielders behind the center forward. Like soccer teams on a playing field after the whistle has blown, both sides were at a standstill. Whatever had happened was over, and the wildebeests were still in the game.

Remembering to be careful lest some predator who had watched us go up the ladder was waiting for us to come down, we left the tower, got into our van, and started slowly for the camp. At the first bend of the track, in a patch of heavy sand, the headlights shone on two full-grown but still adolescent male lions sitting very still and very straight, intently watching the place where the disturbance had been. Not wanting to alarm them, I dimmed the lights as the van labored slowly around them. They turned and looked at us.

Then, suddenly, to our astonishment, they launched themselves at us and, tails high, began to chase us, one on each side of the van. Bounding along like two huge dogs, they seemed to be snatching at the tires. Fearing that we would soon be wobbling feebly on the rims while two adolescent lions tried to pull us through the windows, I floored it. The struggling van lurched forward, the lions fell behind, and in the rearview mirror I saw the distance widening. In my last glimpse of them they were standing in the road, somewhat crestfallen but still much interested, watching us go.

I was flabbergasted. I, who had spent so many nights sleeping on the open ground in the Kalahari and so many more nights lost in the bush in Uganda, had never dreamed of being chased by lions. No such thing would ever have happened at /Gautscha. The Ju/wasi would not for one moment have tolerated being chased or played with or harassed in any way by lions—not in the past and not today.

Then I saw how stupid I had been, and how deeply I had misunderstood Gobaub and Etosha. The beautiful, dry country, the white grass, the clear sky, and the sight every evening of the setting red-ball sun had all misled me. I had been seeing everything as if I were still in /Gautscha thirty years earlier. But it wasn't the same at all. The animals of Etosha didn't know people. Perhaps never before had that population of animals known so little about people. But the people who would have taught the animals were gone.

The people were gone, and the old way was finished. If the authorities had decided that the park would be more natural without lions and had removed them, their absence would not have been more glaring than the absence of human hunter-gatherers after half a million years. As soon as I realized that, I no longer saw the similarities between Gobaub and /Gautscha but, rather, the differences. At /Gautscha, time and rainfall had managed the ecosystem; in Etosha, as we soon learned, the ecosystem was managed so thoroughly by the Department of Nature Conservation that the place sometimes seemed like a farm. The populations of animals were continuously monitored and controlled. The lion population was controlled to some extent by means of long-lasting contraceptive devices implanted under the skin of some of the lionesses. Those lionesses had grown old without having offspring, and after the drug wore off they might be too old to raise and educate offspring. If so, the experience of these lionesses would be lost—a situation more serious than might at first appear, for a lion needs to know a great deal to be able to meet the challenges offered by the environment, and especially the challenges, such as serious droughts, that happen only rarely. In such circumstances, the guidance of an experienced lion can mean the difference between life and death, not only at the time but for lions of the future, who, like links in a chain, will someday also benefit from and carry imparted information. For learning to take place, informed and uninformed lions must share an experience—for learning to take place about human beings, an inexperienced lion would have to watch the reaction of an experienced lion when a human being appeared. If the chain breaks—if there are no human beings to provide the occasion for instruction—the experienced lion has no way to pass on the information, and the information is lost. This important truth about the cat family (and indeed all animals who learn from one another) can be seen in housecats, if the human observer compares the overexcitement of adolescent kittens testing a new prey item on their own with the eager yet respectful manner in which the same kittens keep back and out of the way while an experienced mother cat shows them how it's managed.

In the case of the Etosha lions, it is impossible to guess the age of the imparted information that has now been lost. But a population of lions on the same land for generation after generation could have been passing information that might have been exceedingly old—hundreds or even thousands of years.

Much of the park's intense management was directed toward research, with the animals, of course, its subjects. In keeping with the hard-science trend in behavior biology, the park, which has in the past permitted invasive research on the animals, generally disapproved of unstructured observation, considering the results not measurable and therefore not worthwhile. Katy's work required complex equipment and had a "hard" aspect, in that we collected measurable sounds, and this contributed to our being allowed to work there. We were ordered never to name the elephants we studied, lest we appear sentimental; we were told to number them instead—a strange conceit along the line of white lab coats for scientists, evidently conjured to make scientists appear to be more scientific. Wild animals have no names, it is true, but they certainly have no numbers either. With this in mind we named the elephants anyway, despite the warnings, because names are much easier to remember than numbers.

On the shortwave radio that we had been ordered to keep on so that we would be in constant contact with the park authorities we couldn't help hearing, day and night, the doings of the park biologists as they darted, biopsied, branded, and tagged the game—especially the lions. If under the old way the lions and the hunter-gatherers had kept their distance from each other, under park management the people took such an interest in the lions that any lion could expect to be physically invaded sooner or later by a diagnostic procedure or a telemetric device. For the lions, that meant sudden, probably bewildering intrusions: a drug experience almost certainly followed by pain—the lingering burn of a brand, perhaps (not that pain seems to be much of a problem to lions). Most lions experienced such handling only rarely and hence had little need, or even incentive, to adjust. So the presence of the park personnel, though it changed the lions' lives, probably didn't make a difference that lions would recognize. Nothing that I could detect in the lions' attitudes suggested that they gave much thought to the presence of the park personnel.

What I did detect in the lions' attitudes was that they had no concept of the hunter-gatherers. After we moved our camp to a second, remote waterhole called Dungari, we found ourselves under intensive observation by the Dungari lions, who apparently didn't know what we were but wanted to find out. Like mice in a cattery, we could do nothing without first checking to see whether we were safe from them. Three of us—Katy, her daughter Holly, and I—first visited Dungari to learn if elephants used the water, which at Dungari was provided by the park in a large concrete trough kept full by a solar pump. Dungari was just a place in the woods, really—not a great flat plain with a natural well like /Gautscha or Gobaub. Nevertheless, lions were there, as we began to notice after dark, when, in the van, we started a twenty-four-hour vigil. With me in the driver's seat, Katy in the passenger's seat, and Holly in the back seat, we thought we were watching carefully and seeing everything, so we were more than surprised when the face of a lioness suddenly appeared by Katy's shoulder, framed in the right front window. How had so big a creature managed to creep up on us?

A lion had come with her, we soon learned, but of the two she was the more curious. He was more or less hiding and seemed to be waiting for her to do something. What that might be we weren't sure. We suspected that she might be hunting us, though; it seemed that every time we looked out the window her head would pop up beside the van.

Of course, after we realized she was around and was probably hunting us, I kept trying to find her with the flashlight. I picked up the eyeshine of many other animals, and I often spotted her large, tawny form near at hand, but very seldom could I spot her blazing green eyes. At first, I couldn't understand why. But then it came to me that she might be concealing her eyeshine by averting her eyes. Because she was usually facing us when I spotted her, I began to think she was catching us in her peripheral vision and consciously avoiding a direct look. But why? Do lions know that their eyes shine? Lions' eyes shine in moonlight and even in starlight. Moonlit eyeshine can be by far the most conspicuous feature of a lion at night, especially of a lion hiding in the grass. If a lion should see another lion's eyes shining, could it then infer that its own eyes might also shine? The fact is, we haven't the slightest idea whether lions could make such a deduction.

My own feeling is that they can and do, because cats know so much about eyes, eyesight, and visual perception, as the following observation of a housecat demonstrates. One evening, our son, Ramsay, was sitting on his livingroom sofa when he noticed his cat Buster on the far side of the room, creeping along the wall. Because Buster sprays at night, he is put outside, but because he sprays to claim for himself the parts of the house that he fears will fall to his rival, the family's other male cat, Eddy, he wants to stay inside where he can stand up to Eddy. Therefore, when darkness gathers, to avoid being put outdoors Buster tries to elude the human members of the family. So Buster was creeping along because he had seen Ramsay and was trying to cross the room without being noticed.

At one point, an armchair in the middle of the room blocked Ramsay's view of the cat, and at that point, the cat vanished. Because the armchair was in the middle of the room and the cat had been ten feet away against the wall, Ramsay grew puzzled. Where was the cat? Very quietly Ramsay stood up so that he could see over the chair, and to his surprise saw the cat crouched against the wall, ears low, craning his neck to peer in Ramsay's direction, like a cat in hiding, peeking around a corner. And it came to Ramsay that the cat was indeed hiding from him, using the distant armchair as a screen.

The episode shows very clearly that the cat knew where his body was and how it could be hidden, and had calculated Ramsay's line of sight accurately and from afar without benefit of eye contact. A person in a comparable situation would have felt exposed.

The lions began a more intensive investigation of us when we came to stay at the Dungari waterhole, and they were later joined by two other lionesses. I was quite moved to realize that their investigation of us resembled ours of them. At night, through the fence of a horse camp that the park authorities had urged us to use, the lions watched us sleep, just as during the day we watched them sleep. As we examined their sign, they examined ours, following our tracks to our various latrines, which they unearthed. Our sign meant something to them: frequently, they left their scats beside ours and squirted their marks over the traces of our urine. In other ways, too, they seemed to do what we did. One day, I was sitting near the solar pump at Dungari watching a lioness who lay by the water. She was near enough that I would need to get to safety if she stood up, so I was trying to be careful and not forget her. She was watching me in an equally casual manner. But the day was warm and the air was soft, so it was hard to sustain anxiety. In time, I yawned. To my amazement, without taking her eyes off me she also yawned. Was it a coincidence, her enormous red gape? Was it empathy? Fascinated, I deliberately yawned again. She yawned again! I yawned again and again. But I had done it too quickly. She simply watched me through half-shut eyes. I waited two or three minutes and then yawned once more. She yawned right away. More than excited, I called the other people so that they could see what was happening. One last time, the lioness obliged us with an empathetic yawn. Then, seeing that several of us had gathered to stare at her, she suddenly seemed to get self-conscious; looking irritated, she got up and left.*

Perhaps this lioness had been studying me. Not long afterward she stalked me. Once, when I didn't think she was around, I walked several hundred yards from the safety of our study area to collect some equipment. As I circled through the heavy bush, someone in the study area noticed the lioness creeping stealthily toward me and called to me to come back. So I did, remem-

---

* She was not the only large felid to engage in empathetic yawning. A young white circus tigress named Taji in New Jersey did the same thing in response to an accidental yawn from me. I yawned, she yawned, we paused, time passed, then she yawned hugely and watched me closely, perhaps wondering if I'd keep it up. For some reason I didn't, but I wish I had.

bering to walk as the Ju/wasi had taught me, and here I am to recommend the method.

If she had caught me, would she have killed me? Quite possibly, but not necessarily for food. Cats after all learn much about their future prey by playing with the first few specimens they acquire. They learn how much stamina their victims have, whether they bite, and how fast they move, among other interesting facts. So to pursue her observation of our kind, the lioness probably would have played with me, which could have injured me pretty badly if it didn't kill me.

That same season Katy had an experience that must be unique among field biologists who study large animals. It happened at night in a tourist area of Etosha, near a waterhole that is illuminated by floodlights like the stage in a theater, where in reverent silence the tourists watch the comings and goings of the game. On one side is an area where visiting researchers are allowed to make their camps, and there Katy had put her sleeping bag right next to a length of fence made of lightweight wire. Katy is one of the most sensitive people I have ever known, as fully attuned to the natural world as Ju/wasi. In the middle of the night she woke and looked up into the face of a huge maned lion, who was sitting on his haunches, intently looking down at her. He was on the far side of the fence, but right up against it, and could easily have jumped over it or gone around the end of it or broken through it. He could also have walked through a large gap nearby, where two sections of fence had come unjoined. He looked and looked at Katy. She looked and looked at him, hearing the wet noises of him swallowing his saliva and settling his tongue. He was thinking of eating. Cats are famous for their patience—the big lion watched Katy while the moon slowly rose behind him, shone through his golden mane which the wind was stirring, slowly climbed to the top of the sky, and shone down on both of them. All this time Katy held his gaze and lay perfectly still. The lion continued to think of eating. Eventually, he drooled. Moonlit strings of saliva stretched from his tongue to the ground. Still Katy lay motionless. The sky began to get light. For the first time, the lion looked away. Then

he looked back at Katy, then he slowly stood up and turned sideways. Stiff from sitting so long in one place, he stretched. Then
he walked away.*

Did the fact that Katy held the lion's gaze save her? Possibly so. Eye contact has much meaning to the cat tribe, as
foresters in the Indian Sundarbans discovered through efforts
to protect local woodcutters and honey hunters from the Sundarbans' notorious man-eaters. The foresters found that the
tigers wouldn't attack a person wearing a mask on the back of
his head. Ordinarily, the tigers would attack from behind, leaping on the person's back and seizing him with a death bite to
the nape of the neck. But no tiger would attack head-on. Thus
the staring eyes of the mask inhibited the tigers, some of whom
followed the woodcutters and honey hunters anyway, snarling
at them as if angered to be cheated of their prey. According to
the science writer Sy Montgomery, who has done some astonishing research in the Sundarbans,† the tigers soon learned that
the masks were an illusion and began to attack people anyway.

Katy's interpretation of her experience is illuminating, or so
it seems to me. "I kept trying not to move, but of course I knew
the lion must have been aware of my breathing," she says. "I got
cramps all over from trying to hold still and knew I wasn't suc-
ceeding. So I knew that if I survived, it wouldn't be because the
lion was deceived but because he understood that I respected
him."

Was he the same lion who, at the age of twelve in 1993,
killed and ate a tourist? At the same waterhole, in the floodlit
tourist area, a young man from Germany had, like Katy, been
sleeping on the ground when a lion and a lioness killed and
ate him. According to statements issued by the park authorities, the incident was the first of its kind, and in a way it may
have been, since the young man was probably the first *tourist*
to be killed by lions. But according to information given to me
in the 1980s, he was not the first person to be attacked or

---

* Personal communication from Katy.
† Sy Montgomery, *Spell of the Tiger* (Boston: Houghton Mifflin, in press).

killed, by any means. In 1986 or '87, a SWAPO guerrilla trying to penetrate Namibia from Angola was allegedly killed by
lions and may not have been the only guerrilla to lose his life
in this manner. (Another possibility is that he was killed
by people, maybe by the police in an effort to extract information from him, and the killing was blamed on lions.) Whatever happened, the lions were generally credited with main-
taining the park as a military-free zone, although that, too, is
hard to believe, since SWAPO guerrillas would have carried
weapons. The other lion victims of Etosha were park employees who for one reason or another had had to travel on
foot through the dangerous bush. Foot travel is strictly forbidden to everyone, to park personnel as well as to tourists.
Tourists are enjoined to stay in their cars no matter what happens to them, so any foot travel is undertaken only by park
personnel and only on the rare occasions when their vehicles
break down.

The absence of the hunter-gatherers from Etosha showed in
the habits of lions as well as in their obvious unfamiliarity with
people. Many a waterhole in the park, for instance, was the
headquarters of a pride of lions. The Etosha lions had the mis-
leading reputation of being waterhole hunters, with the impli-
cation that they needed only to wait for their prey to come for
a drink. This wasn't so, of course; never stupid, the prey animals don't drink where they can't see, don't drink at night, and
would drink elsewhere or forgo drinking if they had reason to
think that lions would seize them at the water. No, the lions of
Etosha had to hunt the hard way, like lions everywhere. So no
one really knows why they liked to stay near the waterholes. Yet
lions are excellent observers, and observation is important to
them—hence their empathy. Other cats learn by practicing; for
instance, the hunting skills of kittens of cheetahs and housecats
are sharpened by their mothers, who bring them live animals
to kill. But lions learn mainly by following their mothers, aunts,
and grandmothers and watching how it's done. The warthog
lesson observed by my brother was an exception. Most of the

animals hunted by lions are too big to be transported and too dangerous to be released for the cubs.

So from childhood onward, observation is a useful tool for lions, as it is for all the cats, and perhaps explains why lions station themselves at waterholes, where they can't help but observe the herds that pass by. Perhaps they preselect victims. They certainly get to know the animals individually; their techniques of observation are not that different from the techniques of field biologists whose job is also to know individuals well. Perhaps, however, lions merely note the general condition of the herd and its direction of travel. Anyway, whether or not the practice of waterhole viewing helps lions with hunting, the practice of sitting near a waterhole means that lions don't have to walk far for a drink. Even this could mean a lot to lions, since, like most animals, they suffer from dehydration and need water when they exercise in hot weather.

That's why lions mostly stay still by day. But they can't always. Once near the waterhole at Gobaub we watched a lion and a lioness having sex. Hour after hour they went at it, experiencing one climax after another that would wring from the lion a touching, high-pitched *meow*. The lioness, though, would bellow out a thundering roar, then would twist herself around and smash the lion in the face with the full strength of her powerful arm. As her blow snapped his head to the side he'd shut his eyes, grimacing. After several hours and well over fifty climaxes (I was keeping score but had to stop because elephants approached and we needed to collect some data) both lions got up, walked side by side down to the water, and refreshed themselves with a drink. Although he started drinking before she did, by the time they finished, their tongues were touching the water synchronously. Then they walked together back to their spot, where she raised her hips, frowning; he mounted, thrust, and *meowed;* she roared, turned, and hit him, and they were off again.

The truly interesting thing about waterhole viewing, as I saw it, was not so much that the Etosha lions did it but that the /Gautscha lions hadn't done it. Just as at /Gautscha the presence of the hunter-gatherers' camps appeared to keep the animals at a distance from the water, so in Etosha the absence of people's camps may have allowed the animals to come near. From the archaeological sites that the young official had forbidden us to touch I learned that the hunter-gatherers of Etosha had favored the same kinds of places as the hunter-gatherers of the /Gautscha area. That was perhaps not surprising. What was surprising was that those sites were also favored by the lions. Sometimes the lions of Gobaub chose the very spots where people once had lived. Sometimes the lions chose spots that people would have chosen. In fact, I soon found that as often as not I could locate the lions of Gobaub, even in that vast space, merely by looking around for places that I felt the /Gautscha people of the old days would have liked. I would search with my field glasses for a shady place without too many stones and screened from view on the raised ground around the edge of the plain. There were many such places near Gobaub, and in most of them, at one time or another, we might see a pair of round ears above the grass or spot a sleeping lion. What did it mean? It meant, I think, that the lions had taken over those good places soon after the people had gone.

Seeing that, I felt that I was looking through a window at the distant past. Long ago, in the northern hemisphere, cave lions would have used the same good caves that their human hunter-gatherer rivals used and would have competed for these caves in much the same way as the Gobaub lions competed for the good campsites with the Bushmen. Good caves are scarce, and the lions would have needed them as much as the human beings, for exactly the same reason—any mammal with altricial young in the far north during glacial times needed meaningful shelter, a place where helpless infants could be kept warm. To be sure, the campsites of Gobaub were not as essential as the caves of the north—life for either species could have gone on in less attractive campsites. But cats and people have similar tastes in residences, or cats perceive the good qualities of our residences, so that all over the world there are many examples of cats trying to share with us. The humble female puma of Malta, Idaho, for instance, brought her three surviving children to a long-abandoned farmhouse and took shelter in the empty kitchen. And in India today, the palatial hunting lodges at Ranthambhore built centuries ago by the maharajahs of Jaipur have since changed ownership. Abandoned by their princely owners and overgrown by jungle, these palaces now lodge certain tigers, the original emperors of those forests, who have taken the shady palaces as their residences and stretch out on the cool stone floors.

Long ago, around the southern shores of the Mediterranean, little African wildcats took shelter in people's dwelling places, probably finding the supply of mice and rats and the escape from heavy rains much to their liking. There they stayed. Perhaps they even liked the warmth of people's fires. The earliest cat known is from Jericho (now Israel) nine thousand years ago when one of the few amenities that people had that might attract a cat was fire.

I saw an example of this attraction in 1989 in Bushmanland after I had been there quite a long time and thought I knew most of the creatures of the area, certainly the domestic animals—the cattle and dogs, the horse and the donkey that had become part of Bushman husbandry. I visited the Bushman village late one cold night and was amazed to see a black-and-white housecat crouching by someone's campfire. The Bushmen don't keep pets, and forty miles of empty veldt separated the village at /Gautscha from any other human habitation. So who was this cat? And where was he from? No one knew the answers. He belonged to no one; he was feral. Few people ever even saw him, except on those rare occasions when he crept out of the night to get warm.

My heart went out to the Etosha lions, although their size and might, combined with their naiveté about our species, could be frightening. Once, I was charged by a lioness, and then I felt in awe of the hunter-gatherers who, so long ago, had commanded the respect of lions. I was charged while doing just what the Ju/wasi would have done to move a lioness; she had been sitting right beside some of our equipment, which I had come in our van to collect, and when she wouldn't move I got out and slowly picked up a pebble. Then, speaking very respectfully because I could already see displeasure in her eyes, I gently tossed the pebble as I asked her to please leave. *Whap!* She charged! In the blink of an eye she had covered the distance between us. Luckily I had gotten out of the driver's door, which slammed, rather than the side door, which slid. How would the early hunter-gatherers have wrested respect from an animal like this? And without vans to jump into? If the lions of long ago had in any way resembled the lions of today, their respect would have been hard to come by. The lions in Etosha seemed to respect nothing but other lions.

And of course, elephants. The Etosha lions kept away from elephants much as the /Gautscha lions had once kept away from the Ju/wasi. In Etosha, if lions and elephants met, the lions became more than respectful—or most of them did. One night, one of our group saw an adolescent male lion, tail high, rushing an elephant, but the lion probably wasn't entirely serious, because he gave up quickly. In an East African lion story—the personal communication of a reliable witness—a certain lion once crouched down in the grass to hide from elephants coming from his right. Apparently, he hadn't seen that they were on their way to greet other elephants, coming from his left. Before he could decide what to do, he was surrounded. He then threatened an elephant to try to make her back off and give him a way out of the closing circle. All the elephants were startled to see a lion in their midst, and they roared, screamed, and threatened him. In the excitement, he felt forced to attack, so, leaping at the nearest elephant, he clung to her head. She plucked him off, dashed him to the ground, and killed him—the only possible outcome, really, of a conflict involving a lion and an elephant.

I saw only one encounter between the Gobaub lions and an elephant. The encounter ended very quickly and as encounters go was minimal—a nonevent, really. Yet it stayed with me. Probably I would have learned nothing at all from it if I hadn't first seen the same lions in the same place with a rhino. The rhino was a rather belligerent female, who, with her large child at her heels, often came to drink soon after dark. One moonlit night when the lions were relaxing in the open near the runoff, the rhino seemed to take exception to their presence and she charged. The lions seemed hardly to notice. To my amazement, they did nothing at all until the rhino was almost on top of them, and then, very casually, they got to their feet and, with unbelievable aplomb, moved gracefully toward her, stepping aside at the very last moment to let her charge through. As soon as she was among them, they seemed to flow around her like water around the prow of a boat, to reassemble behind her armored rump. Seeming not to know what had happened, she cantered on for a while before she realized that no one was there. The lions barely glanced at her, as if they had hardly so much as a passing thought for her. They looked, in fact, as if they already knew much about this rhino, as if they had developed their coordinated, dancelike tactic just to avoid her and had practiced the maneuver many times before.

In contrast was the encounter between these lions and an elephant. One evening soon after the lions had been charged by the rhino, they were lying in the same place, a pile of tan bodies behind a fallen log which hid them from the plain. I was watching some of them peer over the log at a zebra who was considering drinking from the runoff when I saw them stiffen, then get up and move apart. Far away, elephants had appeared at the edge of the trees. It seemed to me that the lions recognized these particular elephants. A big adolescent male elephant, about sixteen or seventeen years old, left the others and strode toward the lions with his head high, his ears wide, his tail and trunk up. Although he was at least fifty yards from them, the uneasy lions were watching him intently. For just a moment, the maned lion stood his ground. Then, with his legs braced and his head high, he gave a roar. The elephant answered with a roar of his own. The lion roared once more, which brought the elephant onward at a run. This was more than enough for the lions. Without a sound, they turned tail, scattered like a flock of sparrows, and vanished. In the same way—if not, perhaps, as quickly—would the Kalahari lions once have disappeared before an advance of the Ju/wasi.

Today, even the Bushman-
land lions don't give
ground to people in the
same way as before. The respectful behavior once displayed by
the people toward the lions cannot be depended upon either.
This is surely because the old way is fading in Bushmanland, and
neither the lions nor the people are sure of their positions any-
more. Typical of the new way was a recent photo in a Windhoek
newspaper showing two white people, a lion biologist/conser-
vation officer, and his most recent girlfriend, posing astride a
dead or anesthetized Bushmanland lion under the caption "The
wild pair."

Today, after a wrenching period of transition from their former lives as hunter-gatherers, many of the Ju/wasi are trying to
live as subsistence farmers in permanent villages with cattle.
Such a village, with about sixty cattle, stands on the eastern rim
of /Gautscha Pan, above the waterhole. In the veldt around this
village, the lions are trying to live too. To do so, they seem to
be changing their culture.

The /Gautscha people became aware of this one dark
night—oddly, at about the same time I was charged by the lioness in Etosha. On that night, two Ju/wa men and my brother,
John, who since 1978 had been spending most of his time liv-

ing at /Gautscha to help establish Ju/wa farming, were returning to the village in a pickup when the headlights caught the
eyeshine of ten or fifteen lions in a patch of heavy thornbush
just across the road from the cattle pen. Not liking the look of
such a large group of lions so near the cattle, my brother
stopped the pickup and everyone got out. As in the old days, he
and the Ju/wa men moved slowly and purposefully toward the
crouching, staring lions, thinking to get them up and moving.

However, these lions didn't react in the old way. Instead,
one—a lioness—began to roar. But rather than give a deafening shout, as if to say "Back off!" she began a very loud, prolonged, classical series of roars, such as one might hear from a
lion at night, far away—something like this: *Uuuuuuuwaugh.
Uuuuuuuuuuuuuoooooooooooowaugh. Uuuuuuuuuooooooooh! Uuuuu-
uuuooooooooong! Uuuuoooowaugh! Uuuoo, uaooo, uaooh, ooogh,
ooogh, aangh, unh, unh*. But the most interesting thing was that
just before the earsplitting climax, as the roars rose to a series
of deafening bellows and earth-shaking hollers, other lions
joined the first. Many lions roared in chorus. They were like
the lions at the Kavango cattle post, whose choral roaring had
so badly frightened my mother and me many years before.
Needless to say, my brother and the other men stopped in their
tracks. In the terrifying, echoing silence that followed, the men
stood motionless. Suddenly, the roaring began again, first one
voice, then a chorus. Again came a profound silence, and again
the roars. Four times, one lion started the roaring and other lions joined the chorus while the men stood absolutely still. After the fourth chorus, the silence lasted so long that the men
began to inch back toward the pickup. When they got inside
and turned on the headlights, the lions were gone.

Why were the modern /Gautscha lions so different from
their predecessors? To loiter at the village as if they didn't care
who saw them, and not to move off at the approach of the men,
showed that they had changed their ways. Yet the choice of
meaningful vocalizations instead of physical violence showed
that they were forming customs of their own, different from
those of the Etosha lions. Why?

The choral roaring could have been intended to show the size and solidarity of their group, which a listener would ascertain from the volume and the number of voices. A massive choral assault might be useful when two groups of lions meet if the ownership of a resource is in question. Almost certainly, the /Gautscha lions meant to show my brother and the Ju/wa men that the human claim on the area, with its resident cattle, would no longer go unchallenged. In contrast to the silent Uganda lions, who showed their deference to human territorial claims by refraining from roaring, as if they were fugitives on the land of other lions, the /Gautscha lions made it clear they would not leave without protest just because some people wanted them to go. Yet the old days weren't quite forgotten; evidently the lions had enough feeling for the ancient truce not to savage the men, as any of the Etosha lions almost certainly would have done.

But the Etosha lions were in a stable situation—at least from their point of view. For them, the past was lost. In contrast, the /Gautscha lions were in a period of transition. The old way must have lingered in the memories of the older lions, memories that they seemed to use when trying to cope with the violent economic and social changes that threatened to overtake them.

Since the late 1960s, in the eastern section of Namibia's Kalahari, the area now called Bushmanland, environmental productivity had been diminishing. By the seventies, Bushmanland had lost two of its three permanent waterholes to Bantu pastoralists and had lost its western territory to farms. In the remaining portions, overgrazing by livestock, ruts made by vehicles, and the prevention of fire had reduced the vegetation in some places to thornbushes on hard sand. Legions of South African hunters with automatic rifles and all-terrain vehicles had all but eliminated the wild populations of many water-independent grazing animals to make jerked meat, or biltong. Ghemsboks had become rare, and elands had virtually vanished. Factories to process the meat and skins had become a big business around the Kalahari. And hundreds of miles of fencing had prevented thousands of grazing animals—giraffes and wildebeests in particular—from reaching their sources of seasonal water. They died trying, and their dry corpses lie beside the fence.

After our first season of elephant research in Etosha, I went to /Gautscha. Unlike the rest of my family, who had continued to work there, I hadn't been back for thirty years. I came alone from Windhoek in a rented pickup, a drive of three hundred miles, and stopped for gas at the only gas pump in Bushmanland. It stands at the edge of the Kalahari, before the track leads to the interior from the administrative center called Tsumkwe. But the gas pump was closed, so I had to wait, and as I waited, I looked around.

Once Tsumkwe was just a tree in the middle of an endless wilderness, a huge baobab that still dominates the town. Almost forty years earlier, my father and his friend Claude McIntyre had made the track that led there from the last place within the farming area, the town of Grootfontein. They had mapped the area, and they had dug a well. Since then, the current administrator had built his house in the shadow of the tree and had capped the well, thus dominating the only evident source of water in many miles of hot, thornbush savannah. Remembering the water from long ago, and noticing a faucet where the well had been, I got a cup from my pack and went to drink from it. But evidently there were rules about that faucet. A German guest of the administrator's emerged from the house to tell me that the faucet was private and that I couldn't draw water there. Very surprised to be refused water in that arid climate, under the blazing sun, from a well dug by my own father, I left, still thirsty.

It gave me a strange feeling to stand alone in the middle of the road, part of which I myself, at the age of nineteen, had pioneered in a jeep. Now, a few widely spaced, boxlike houses stood far back from the extra-wide road, which went straight for the horizon where the wind stirred dust devils. Thornbush and sand were what the herds of government livestock had made of that segment of the Kalahari. The barren landscape was hung with dust.

As I waited, a few people passed by. All of them looked at me. Some were young Ju/wa people on foot; others were white people in vehicles. Of course, no one knew me—in fact, some of the whites looked at me with considerable suspicion. And perhaps not unreasonably—I was a gray-haired woman alone, wearing dusty clothes, seated upon what passes these days for a running board, and there was no reasonable explanation for my presence. No one spoke to me, although every Ju/wa person met my eyes, serenely checked my face to see if it held racial rudeness, and upon finding my expression acceptable, nodded kindly. Apparently the administration had not been able to erase the intense, humanitarian courtesy that the Bushmen had practiced in the old days. In contrast, the whites—or my own people, so to speak—stared scornfully once they realized that I was a stranger.

At last the hour of siesta came to an end and the storekeeper returned to the store and opened the pump. As he filled the tank of my pickup, I heard a scraping sound and noticed in the distance a Bushman man, dressed in a coverall, beginning to rake the yard of one of the administration buildings. Every few minutes he would stop raking and look at me intently. And then, as I watched, he let the rake fall and began to walk toward me. On and on he came, not following the road but walking straight across the intervening yards, paying so little attention to the white government's innovations that he passed right between two of the whites themselves—administrative personnel who were waiting near their vehicle for a turn at the gas pump. They might have been a pair of posts. A few feet away from me, the Ju/wa man stopped, looked me right in the face and said, "You're Di!ai."

And I was. Thirty-seven years before, Di!ai had been the name the Ju/wasi had given me. But who was the man? And how had he recognized me at all, let alone from such a distance and over all the years? I had had no way of telling people I was coming, so it wasn't that anyone had been expecting me.

What little of the language I had ever known had more or less deserted me, but I began to grope for words. Realizing this, and seeing my amazement, he laughed. Then, very carefully, in a sort of pidgin baby talk suitable to my incapacity, he told me his name and his parents' names, and, after racking my brain to put all this together, I realized that when this man and I had last seen each other, he had been a little child.

Ever since I had been in the Kalahari, I had thought about the Ju/wasi I had known, remembering them, speaking of them, often dreaming about them, always wondering what this or that individual might be doing and what was happening in their world. But never once had I thought about the very young children. In those years, before my maternal instincts had rightly gotten started, children under five or six seemed devoid of interest and were little more than scenery to me. What an injustice I had done the Ju/wa babies! I laughed. Then I cried, probably because I had learned so much about life in the intervening years but evidently not everything. Then I laughed again. The man waited, smiling. I told him where I was going. He said he'd go with me.

Remembering the rake, I said I'd wait for him to finish what he had been doing. He said that wouldn't be necessary. So we got into the pickup and banged the doors and together drove to /Gautscha.

A large number of elephants had come to live in Bushmanland, where before there had been none. Virtually all were males between the ages of fourteen and fifty. Most of them seemed to know one another, and many had bullet wounds in their bodies or bullet holes in their ears. No one—neither the Ju/wasi nor the government's conservation officers—knew who these elephants were or where they had come from, since no elephants had ever been more than transient, seasonal visitors to Bushmanland before. In my opinion, the bullet holes suggested that the elephants were refugees from war, culls, or poaching, and the fact that very few females or young were with them suggested that they had made a difficult journey from very far away. Even more moving and more interesting, if possible, is that they materialized in Bushmanland very inconspicuously. Elephants had not been seen roaming at random through farms and towns, searching aimlessly for somewhere to stay. These elephants had known about Bushmanland and had zeroed in on it. How so many of them knew where to go is less clear.

Bushmanland was not a haven for them, no matter how far they had come or how difficult the journey had been. Their presence quickly attracted a repulsive hoard of millionaire hunters, mostly men from North and South America and Europe who were eager to pay immense sums of money to the Department of Nature Conservation, which, incredible as it seems considering the endangered status of elephants, sells licenses to rich sportsmen for the pleasure of killing the elephants. The department encouraged these hunters to buy licenses to kill as many different kinds of animals as possible, so these sportsmen shot lions too.

Loss of habitat, shortages of food, and being hunted were some of the general problems faced by all the lions of Bushmanland. However, the /Gautscha lions also had to cope with a change in their water supply. The waterhole was never meant to supply the sixty cattle that by 1982 were in /Gautscha. Two years later, halfway through a severe eight-year drought, the waterhole was dug out in an effort to increase the water supply. After that, a deep well was drilled at a distance, fitted with a pipe, a pump, and a trough. The new well was accessible only to human beings. Meanwhile, elephants began to haunt the waterhole—by the time I visited, six or seven huge male elephants would be at the edge of the pan every evening, waiting for dark before moving their enormous, conspicuous selves out of what little concealment the bushes gave them and coming to the water to drink. But the cattle, who drank from the waterhole in the afternoon, would have used all the water, and nothing but mud was left. So the huge, thirsty elephants, who needed about forty gallons of water apiece, would stand all night by the waterhole taking little sips as the water oozed in. It hurt to watch their subdued desperation. And it hurt to see that not much water was left for any other creatures.

Nor was it safe to approach when the elephants were drinking. At the time of my visit, two huge male elephants, one in musth, also haunted the pump by the cattle trough, trying to work out some way to get water from it. One night a young man, a foreigner who had been working with the Bushmen, thought to chase them off. Knowing how dangerous a musth elephant can be, I and other people of the small community at /Gautscha beseeched the young man to leave the elephants alone, especially since the tents, the vehicles, and the little mud houses, which were the only shelter any of us had, would offer no protection whatever against an angry elephant. But even though all of us were vastly more experienced than this young man, and even though I for one had just participated in a long-term study of elephants and could point out the symptoms of musth exhibited by this stupendous individual, the young man paid no attention to our warnings or appeals. Instead, he got out a police whistle, flounced busily up to the elephants, and blew an annoying blast.

Whatever had made this young man think that an irritating noise from a little primate like himself could frighten away a monstrous, experienced male elephant twice his age, an elephant with bullet holes in his ears, desperate with thirst, and revving on testosterone? Bellowing out a terrible roar, the elephant charged. The young man turned and ran straight toward us.

An elephant runs at thirty miles an hour. A person runs at fifteen miles an hour. The outcome of the chase seemed clear. The only question was where the once arrogant young man would be by the time the elephant caught him. If I had a rifle I know I wouldn't have shot the elephant. I love elephants, and this one wasn't at fault. But I might have been tempted to drop the young man before he led the elephant up to us. From reading Ernest Hemingway's "The Short Happy Life of Francis Macomber" for a course in American literature in college, I had learned that such crimes are not always punished. Anyway, by the time the elephant had finished with the young man, no one would have been able to find the bullet hole.

But the elephant was more merciful than I am. He wanted to frighten the young man, not to kill him. With no more effort than a person makes to swat a fly, he could have pulverized his tormenter, but he didn't. Towering above the running man, he suddenly stopped short, gave a snort, flapped his ears, and then, turning on his heel, he walked back to the pump where he could smell, but not drink, the water.

As for the lions, even in the face of so much change, they seemed to be keeping some of their old customs. Perhaps they were trying to keep the structure of their lives unchanged. More than many other animals, cats value sameness, and possibly lions resemble their little relatives in this preference. Sameness, after all, has enormous value to animals whose lives are close to the edge, whose food is unpredictable, and whose enemies include their own kind. Sameness helps a cat to remember things, and to a certain extent to predict the future. For example, by monitoring a certain cattle boma (corral) while patrolling his territory, a Kenyan leopard figured out the farmer's routine of visiting the boma for the purpose of collecting the calves who had been born during the month and moving them to safer quarters. The farmer habitually did this on the first Tuesday of the month, so the leopard made a point of visiting the boma on the preceding Sunday or Monday, and helping himself to a calf. Thus the leopard not only understood the farmer's complicated routine, but had developed his own routine in response, and by leaving as many days as possible for calves to accumulate in the boma, he minimized his chances of being discovered while maximizing his chances of finding a calf there to eat.*

Housecats, too, value routine even if they realize no direct benefit from it. Our cats come home in the evening—a skill we taught them by rewarding them hugely for compliance. To us, their early homecoming has an advantage—at night our area is dangerous because of owls and coyotes—but the cats aren't afraid and would never come home because of predators. And they no longer come home because of the snacks—careless cat owners that we are, we have gotten used to their regular evening appearance and we usually forget to offer them even a single drop of cream. No, they come home for no better reason than habit. They even nudge me at dusk if I'm working in my office. It's time, they say.

Sameness is security to the cat family, especially if the sameness involves group activity. To our housecats, assembling at

* Mizutani, Fumi, "Home range of leopards and their impact on livestock on Kenyan ranches," in *Mammals as Predators*, N. Dunstone and M. L. Gorman, eds. (Oxford: Oxford Science Publications, 1993), pp. 425–39.

night means drawing together. Once indoors, they disperse, and would eventually go back out again if we let them. But their frustration at being confined all night does not outweigh their satisfaction in keeping the routine.

Like all cats, the /Gautscha lions probably valued sameness, which helps to explain why even after many years, some parts of their lives seemed unchanged. Their group, for instance, numbered thirty, the same as the group long ago. Also, the modern lions apparently rested in some of the old places—shady thickets in the bush southeast of the pan. And the lions still used an old game trail (on which, by the eighties, the tracks of vehicles had been superimposed) to travel between their resting places and the waterhole.

Some of their customs were new, however. For instance, they had apparently changed the size of their hunting parties. Although we didn't really know the size of their parties in the old days because we never saw them, we did find tracks and heard calling and answering that suggested that the lions favored groups of four, five, or six. In the eighties (albeit on somewhat impressionistic evidence), the lions seemed to prefer hunting in pairs. At least, this was so in any sightings. Perhaps the reason for this practice was the available prey. Although most of the large antelopes were gone, there were still kudus. Few were big or had long horns, as in the old days. Most were medium sized or smaller, but there were plenty of them, perhaps even more than before.* Duikers and steenboks were also present in reasonable numbers. But that was about all. Since there was little else to hunt, these small-to-mid-sized antelopes provided almost all of the food of the modern /Gautscha lions, who were thus in a situation similar to that of many tigers. Three lions could eat from a medium-sized kudu doe. Two lions could possibly make do with a steenbok or a large duiker. But if more lions were present, some might go hungry. Better

---

* Unfortunately I know no explanation for the kudu population; perhaps the kudus flourished in the absence of competitors, or perhaps the altered vegetation favored them.

to hunt in smaller groups to take the best advantage of the reasonably large supply of small-to-mid-sized prey.

Housecats are quite capable of changing specialties. In his
early years, our cat Orion specialized in birds and would jump
straight up, six or seven feet into the air, with one paw extended
like an outfielder, to catch them as they flew down to a feeder.
He later stopped hunting birds altogether and for one summer,
as I have mentioned, took up the sport hunting of chipmunks.
Toward the end of his rather short life (we believe he was killed
by a coyote) he had shifted again, this time to mice and voles.
If housecats can change their hunting preferences so readily for
no apparent reason, why couldn't lions, especially lions with an
excellent reason?

If a change in the food supply swayed the lions into adjusting their hunting style, the change in the water supply also
caused new behavior. On the very first night of my visit to
/Gautscha, I saw something I had never seen before: lions at the
/Gautscha waterhole. Just a few hours after dark, six or seven
lions appeared on the ledge of rock above it. But the waterhole
was empty. Two elephants had drained it dry and were waiting
for it to refill. On the ledge, the lions wove back and forth as if
wondering what to do, then left all together, abruptly. I next saw
them at the empty trough by the well where people had pumped
water for the cattle in the afternoon. Although the trough was
empty, the lions at last found water where the village dogs found
water—in the muddy footprints of the cattle. They drank these
dry, then left on the road.

Surely the biggest change in the environment, from a
lion's point of view, was the introduction of cattle. When the
cattle arrived in 1982, the lions, of course, killed some of them.
But they didn't kill very many. For this there is no easy explanation. Although all cats are somewhat traditional in their
hunting preferences, and though the cats seem to have different techniques for different kinds of prey, it simply does not
seem possible that the lions didn't consider hunting the cattle
from the moment they first became aware of them. One whiff
of that dizzying, grassy scent would have set a lion's mouth wa-

tering. Nor were the cattle protected by the people—or, at any rate, not by day. By day, the cattle wandered far from the village under their own recognizance and came home each night of their own accord to a dung-filled pen fenced with a few strands of wire. Even I, in late middle age, could almost have jumped that particular fence; a lion would hardly have noticed it. By rights, considering the size of the pride, the absence of other food, and the cornucopia of opportunities, the lions should have killed a cow every few days. Yet for five years they didn't; they kept scouring the bushes for steenboks and duikers and, except for some very unusual instances, they left the cattle alone.

To me, this restraint was almost incomprehensible. While it is true that cats specialize in prey items, true that the leonine culture of the /Gautscha area wasn't focused on cattle, and true that certain kinds of hunting techniques must be learned, lion populations have nevertheless responded to invasions by new species in far less time than five years. When in 1898 the Kenya Uganda Railway was built between Mombassa and Lake Victoria, certain lions of Tsavo learned to hunt the railroad workers soon after they had bridged the Tsavo River, and showed tremendous ingenuity in obtaining more than sixty of these unfortunate men. As housecats open gates and cupboard doors, the lions even opened the doors of railroad cars at night and crept quietly inside in quest of preselected victims.* Lack of ability is not a characteristic of lions, or indeed, of any of the cats.

And yet, in a way that I couldn't precisely define, the restraint of the lions seemed like the old days. Did the lions recognize the cattle as part of the human domain? A dog would. Are lions, those acute and empathetic observers, who in their own culture recognize certain kinds of ownership, less able than dogs to recognize ownership in other species? Unfortunately,

---

* M. F. Hill, *Permanent Way* (East African Railways and Harbors, Nairobi, Kenya—no copyright notice—c. 1949), pp. 172–74. See also Lt. Col. J. H. Patterson, *The Maneaters of Tsavo* (Macmillan, 1927).

we still know too little about lions (or any other animals) to answer these questions in any meaningful way.

Not everything was mysterious, however. The cattle themselves had much to do with their relative immunity from lion
predation. When the pen was opened in the morning, the two
or three cows who provided leadership wouldn't just start walking, as American cows might do, but instead would stand among
the others for a while, as if waiting for something. Because they
had long since eaten the grass near the village, the cattle might
have been contemplating a number of grassy places a mile or
more away. Eventually, they would choose one of those places
and start walking toward it, always taking several precautions.
At least when I was there observing them, they usually varied
their direction, thus making their whereabouts unpredictable.
They never left before the sun was high and always returned
long before sunset, thus avoiding the times favored by the crepuscular lions. They always traveled in single file, especially
through heavy cover, as infantry soldiers are taught to do in jungle warfare. And finally, they always stuck together. Once, when
a young cow missed leaving with the others, she became pan-
icky until, casting about, she found their scent and ran after
them with her nose to the earth, like a hound.

To the Ju/wasi, accustomed to animals who knew what they
were doing, the enterprising attitude of the cattle seemed to be
the natural order of things—the old way. Might this not also
have been true of the lions? The few times they had hunted cattle, the technique they had used was reminiscent of the past; a
lioness came at night to roar repeatedly outside the pen, which
caused the cattle to panic and lunge at the fence until they had
jumped out or broken through, whereupon one of the lions
killed one of them. The episode reminded some of us of the lioness whose roaring had frightened us long ago, because the
roaring was of the same type—not the classical roaring that rises
in a crescendo and falls away to grunts but a series of similar,
loud, steady roars. I wondered if the lioness really intended to
panic the cattle. Perhaps something else had induced her to
roar. The cattle pen was precisely on the site of our old camp,

and she had arrived from exactly the same place as her predecessor of thirty years before—from the trail that led from the
southeast to the water. She also faced the same way while roaring. Was all this just coincidence? Or could the place itself mean
something? On the plain, it was the high ground. Might that
make it a good place to roar? We didn't know—we just listened
nervously, as we had decades before. In response, the Ju/wasi
built a thick thornbush barricade around the wire pen, but not
with the thorns turned out against the lions, as one might expect. Instead, the Ju/wasi turned the thorns in against the cattle. That was to discourage stampeding and seemed to confirm
the thought that the cattle had only themselves to blame if the
lions caught them.

When I visited /Gautscha after our second season of research in Etosha, an event escalated the cattle killing by the
/Gautscha lions. A new, young, farm-raised bull was brought to
/Gautscha. There he met the enormous resident bull, Boesman
(Bushman)—so named by his Ju/wa owner because of his impressive size and his unruffled demeanor. The new bull couldn't
avoid being squeezed into the pen with Boesman during his first
night. But in the morning, out of respect for Boesman and unaware of the danger of lions, the new bull stayed far behind the
herd. That same morning, a pregnant cow began her labor. Because of the people's laissez-faire attitude, nobody noticed her,
so she, too, went out with the other cattle, and she, too, lagged
behind the herd. The cows who led the herd chose an unusual
direction that morning—southeast, to the area where the lions
sometimes spent the day. There the laboring cow began to give
birth and soon caught the attention of two lionesses. They attacked her. The knowledgeable, experienced cattle left the
scene immediately, but the new bull apparently tried to help the
cow whom the lionesses were killing. The lionesses killed him
too. They ate their fill and then went into the bushes and fell
asleep.

That night, no one noticed at first that the cow was missing. But when the new bull failed to come home the people
began a search. In the morning, they found the partly eaten car-

casses—about half the cow, one little foot of the calf, and most of the bull. The people at once built a fire, cut a lot of the meat, and cooked it for a nourishing feast reminiscent of many such feasts inadvertently provided by lions in the old days. A party atmosphere prevailed. Afterward, the people cut the rest of the meat into strips and carried it home. The lionesses, it turned out, were only about seventy feet away all the while, but, in the old way, they said nothing.

That evening the young man whom the elephant had chased persuaded some of the Ju/wa men to let him drive them to the site. He would provide light so that when the two lionesses returned to look for scraps the men could shoot them from the vehicle. Although the idea was that the Ju/wasi would stand in the open back of the pickup while the young man sat safely in the front seat, the Ju/wasi agreed, since they are fearless (and did not realize that the act would be illegal). But when the lionesses became vaguely aware of the poisoned arrows whispering by, they crept into the bushes uneasily. Later that night, I saw them trying to drink at the waterhole. A few days later, I went home. It was September.

The rest of the story was told to me by my brother, who was present. The rains had been scanty the year before and the drought was felt in October. Then a bad fire to the north, south, and west of /Gautscha Pan burned most of the grass. All the grazing animals were hungry, but the wild ungulates could leave the area and go to search for better grazing. The cattle, however, had to stay at /Gautscha, and they were especially hungry. One morning, a Ju/wa herdsman took the cattle to the southeast, where, perhaps because the cattle had been hesitant to venture there since the killing, some grass was still standing. After delivering his charges, the herdsman went hunting. Soon the cattle attracted the attention of the lions, who that day were all together, thirty strong. Among them, undoubtedly, were the two cattle-killing lionesses, with the memory of their September success fresh in their minds.

The cattle, too, would have remembered the episode. They presumably hadn't wanted to go back there in the first place.

Anxious to leave, they split into two groups. So did the lions, apparently in order to follow the two groups of cattle, which went separate ways. That in itself was somewhat unusual. More unusual was the lions' method of attack. Evidently, instead of concentrating as one team upon killing and eating a single victim, many lions or teams of lions suddenly attacked many cattle simultaneously. Perhaps some lions or teams of lions slaughtered many cattle in the first group and then moved on to the second group, or perhaps each group of lions slaughtered many of the group of cattle it had been following. Whatever happened, the veldt was strewn with corpses—eight together in one area and four more about a mile away. The next day, the people backtracked along the path taken by the surviving cattle and found twenty lions with the group of eight corpses and ten lions with the group of four. My brother, who was concerned about preserving the livelihood of the Ju/wasi, shot and killed two of the lions.

He suggests that the drought, the fires, and the resulting lack of grass made an unusual set of circumstances that caused the wild game to go elsewhere, leaving the lions no choice but to kill cattle. He has no explanation for the method of killing—the massacre—and wonders if it will be repeated. As for the lions, they will long remember the details of their achievement, even though at the time it was apparently a lion excess, an exception to the rule. Indeed, they may try another massacre someday, even though the first cost two of them their lives. After all, they aren't cowards. So far, however, that hasn't happened. Even before the rains came and the grass grew back, the surviving lions returned to their customary abstemiousness. At the time of this writing, they kill cattle only rarely and only by their former desultory methods. No one knows why.

PART THREE

New Ways

Born of the Eocene, cats have picked their delicate way into all continents except Australia, arranging themselves around the world in their fluid, rose-shaped territories, surviving in all climates except at the poles. Creatures of the edge, skilled hunters whose only food is meat they or their mothers kill, most cats are rare wherever they are found. Pressure from the burgeoning human population makes many kinds of cats much more rare; pressure for space crowds them from their homes and a rapidly growing market for their body parts is eliminating many of their species.

This is particularly true of the spotted cats and tigers. Spotted cats such as ocelots, marbled cats, black-footed cats, and margays fall victim to the fur trade, while a recent survey provides the shocking truth that despite the success of Project Tiger—a major campaign mounted two decades ago to save the species—less than four thousand tigers remain alive in the wild outside of the Sundarbans, where the world's last viable tiger population still clings. Most of the other tigers have been slaughtered by poachers and their bones have been ground up and sold as medicine in many Asian countries. Today the human population of these countries is close to two billion, meaning that the Asian buyer/wild tiger ratio is five hundred thousand to one. In other words, for every individual wild tiger from Sumatra to Siberia, there are a half million people who may think they need a piece of him.

One of the most important roles that concerned people can play in tiger conservation is to push hard for sanctions on countries that ignore the illegal trade in tiger parts. Another is to help support the forest guards in reserves where wild tigers are still found. Still another is to help support the captive breeding plans of the Species Survival Commission, a branch of the International Union for the Conservation of Nature (IUCN).

Four thousand tigers are extremely few, even if all belonged to one subspecies, or to one continuous population. But these four thousand represent all the living subspecies of tigers, from the big, northern Altai tigers to the slender, gracile tigers of Sumatra, and most of them live in tiny, relic populations on little reserves scattered here and there like islands. These populations have no communication with one another, and are now almost as threatened by inbreeding as by the Asian trade in tiger parts. Taking a leaf from the book of the farmer, scientist and game management officials are now trying to manipulate breeding through the artificial insemination of wild females in these island populations with semen from captive males, or from wild males of other island populations.

The process is only now being developed and is not without difficulty. Who, after all, dares to extract semen from a wild male tiger? The tiger must be tranquilized and his semen must be extracted by a process known as electroejaculation. His semen must then be transported to a wild female who must also be tranquilized and artificially inseminated. Hopefully, she will own her own ranch so that she is self-supporting, and hopefully she will be capable of teaching her skills to her children so that they, too, can survive on their own when the time comes. The male, on the other hand, can be from anywhere, even a zoo. His subspecies and genetic history are what matter, not his skills.

Tigers whose birth results from electroejaculation are far better than no tigers at all. But in the future, scientists will be making the choices that tigers traditionally made for themselves. Thus, as genetic intervention becomes the way of tiger reproduction, the participants will perhaps not live in cages but neither will they be wild.

Even then, in the foreseeable future, it seems very likely that the majority of the world's tigers will be found in captivity, either in circuses or in zoos. Under almost all conditions, tigers, like housecats, breed easily and well. And here, the problem is eerily similar to that of wild tigers. Zoos and circuses cannot provide homes for all the tiger cubs born annually in captivity. Places are found for the lucky ones. What happens to the rest?

The answer depends largely on the ethics of the zoo's management. Short lives are the fate of many of the tigers who participate in breeding programs for endangered species. As a rule, most breeding tigers are allowed to contribute relatively few kittens to the gene pool of their subspecies. After they have done so, their usefulness is over. They drain the zoo's scarce resources and waste precious cage space, so they are sold to other zoos or killed to make room for new genetic contributors.

Zookeepers would prefer, of course, to find good homes for these tigers, and responsible zookeepers are unwilling to sell them to just anybody, first because in the United States (if not in all countries) the laws pertaining to the private ownership of big cats are, fortunately, very strict, and second, because the zookeepers are well aware of the fate of tigers who fall into the wrong hands. Some unwanted tigers are sold to purveyors of Asian folk medicine who grind their bones into powder. Other tigers are sold for their hides. Still others are sold to game farms where people calling themselves sportsmen slaughter them for excitement. The young tiger is taken in a truck to the scene of his execution, where, as the sportsman waits, his weapon raised, the back of the truck is opened, people with sticks poke the tiger and the frightened animal comes cringing out. *Ak ak!* The tiger lies dying and the would-be sportsman has become a tiger hunter—a regular Jim Corbett in his own eyes.

Rather than encourage a trade in tiger bones and hides or provide easy thrills for lazy or inept sportsmen, many responsible zoo managers feel that euthanasia is a better end for un-wanted tigers, sad as it seems. At least the bodies can then be studied for the benefit of the species, especially for the captive breeding programs. The people who run the breeding pro-grams are well aware of the moral dilemmas stirred by captive breeding and fully realize the consequence to individual tigers of the effort to keep the subspecies alive. In short, in a situation where all solutions to the problem of breeding tigers are fraught with sadness and difficulty, the zoo managers have probably found the least pernicious.

Few people would deny the necessity of zoos. In a world where contact with nature is no longer available to most peo-ple, zoos are the only place where we can have even the most minimal experience with other kinds of creatures. Beyond that, captive breeding programs are the only possible way to keep many endangered species and subspecies on the planet. That we have come to this point in our ongoing ecological disaster is not the fault of the zoos, which should be encouraged and supported, no matter what the consequences may be to some of the individual animals involved.

A fortunate few of the captive-born tigers get to live out their lives. What is the quality of those lives? Some zoos now let their tigers spend their days in large, open, outdoor enclosures. The ambience is therefore more pleasant for the zoogoers, who pre-fer to watch tigers sleeping on grass in naturalistic settings rather than on concrete floors behind bars. How the tigers feel about the difference is less certain. When recently I spent a few days watching a pair of tigers in a major urban zoo, I was impressed with the size and beauty of their pen but was struck by the fact that the two tigers hid all day behind some bushes, resting qui-etly in the only place that could conceal their large bodies. When evening came, they bounded to a hidden door on their hillside and waited eagerly to be let into the tiger barn where, in small cages, they would spend the night. Out of the public view, where no one would shout at them or explode flashbulbs in their eyes, they could communicate with their keepers and with several other tigers—much more pleasant for them than the outdoor enclosure with its ever-present, anonymous crowds. The tigers would enjoy a relaxing but informative night on shelves in their little cages, and in the morning they would again be driven outdoors with threats and shovel banging to face another day of stressful tedium. I found their preference both interesting and touching, especially since it was not food related. The tigers were fed outdoors.

In contrast to the zoo tigers are the circus tigers. Mostly, these are generic tigers who have been bred as most housecats are bred, by a chance meeting of the parents. Hence, as a housecat can be, say, part Siamese, so a generic tiger can be part Siberian. On the other hand, some circus tigers are very fancy indeed. As nothing stops cat breeders from raising hairless cats or cats who flop over as if dead when they are handled, so nothing stops many tiger breeders from breeding tabby tigers or white tigers. The former supposedly have distinctive stripes and the latter are merely an aberrant form in which white replaces the usual orange color, often with the undesirable side effect of an intractable personality and seriously crossed eyes.

White tigers are merely freaks, not an endangered species, as some breeders claim. That a white tiger is (at the time of this writing) the main or only tiger in the National Zoo should not be taken as an endorsement of the phenomorph—her presence is merely a tribute to the zookeepers, who won't dispose of her just because her color is abnormal. They keep her in spite of her color, not because of it. In contrast, two men in Las Vegas have produced what seems to be a pure white tiger with no stripes. It is hard to imagine that such a feat of manipulated inbreeding could be achieved without much trial and error and without the births of many, many kittens. Where are those kittens now? Rumored to have been deprived of his claws and his eyeteeth, the all-white tiger is part of a white-tiger menagerie that is utilized in a nightly spectacle put on by the two men. The eerie abnormality of the creatures and their life on a glittering stage in Vegas bespeak the ultimate degradation of wild animals in the modern world.

However, the personal lives of these animals, or some of

them, are rumored to be rather pleasant. Supposedly the favorite tiger lives in his owner's house and has a room of his own.
Do big cats want to live with people? I have no information that
could answer for the Vegas tigers, but I heard of a lion in similar circumstances—though far less opulent, to be sure.

The lion in question was a large, black-maned Kalahari
male. He, a man, and a male dog—a Great Dane—tried to make
their living by performing throughout the South at local events
such as mall openings. But something had gone wrong. Perhaps
the trio couldn't make ends meet. At any rate, they had been
forced to disband. What then happened to the dog and the lion
is unclear. When I met the man, he was working as a security
guard in Florida at a small circus museum and was very much
alone. An animal lover, he mourned his former companions,
with whom he had shared a trailer. Struck with the picture of
spending a night in such close quarters with a loose lion, I asked
if the lion hadn't had a cage. "Of course he had a cage," the man
answered sadly. "But he preferred to sleep with me and the dog
in the trailer."

What, then, is the best life for a captive lion or tiger? Strange
as it seems, the answer may be that the best life for a large captive cat is that of a circus performer. A fortunate circus tiger, in
my view, might share a cage with another, compatible tiger, in
a collection of ten or twenty fellow tigers whose owners not only
train them and perform with them but in all ways share their
lives. Needless to say, a collection of tigers can seldom be
trusted to caretakers other than the owners, no matter how experienced or well meaning the other caretakers may be. Therefore the owners and the tigers are almost always together. If the
tigers are the prisoners of the owners, the owners are also the
prisoners of the tigers, in a state of communal existence that can
last for many years.

When on the road, both the people and the tigers live under conditions that seem uncomfortable and even severe. The
owners live in small trailers and the tigers live in traveling cages
on wheels, each cage about twice the length of the tiger who in-

habits it. Sometimes nothing better than a large tarp or the edge of a circus tent shelters these little groups of people and tigers, just barely protecting them from wind, sun, and rain. Most of the time the people manage to eat seven days a week and to feed the tigers six days a week (even in the best zoos tigers fast for one day a week—supposedly a day without food is good for them). But, again because of the all-meat diet required by the cat family, life on the road is sometimes marginal—one tiger can eat $300 to $400 worth of meat a week, and the distributors and slaughterhouses from which the owners buy the meat don't always extend credit, certainly not to circus people who may be here today and gone tomorrow. So some days the tigers might not get full rations. Even so, if I were a tiger I wouldn't mind the circus life, even with the long hours and the hardships.

To be sure, the circus wouldn't be my first choice. First, I'd like to be a successful, high-status wild tiger on a large well-stocked ranch of my own in India. But after that, I'd choose the circus, assuming decent treatment. Third, I'd choose to be a less successful wild tiger, perhaps one in an overcrowded community or in a managed island population. My last choice would be the zoo. Even if the zoo were an excellent zoo like the Minnesota Zoo or the Brookfield Zoo in Chicago, which would put it high above the marginal roadside zoos, I'd find the boredom difficult. And the tigers do too.

One of the most moving facts about tigers that I've ever heard comes from Dr. Ronald L. Tilson, biological director of the Minnesota Zoo and senior editor of *Tigers of the World*. During a long-term study of tiger reproduction, essential if tigers are to be saved, numerous tigers lived in a busy laboratory in rather small cages, where they were immobilized regularly so that their blood could be drawn. One might think they would suffer. But they liked the life! Perhaps they didn't like the needles, and perhaps they got angry when tranquilized with darts, but they liked the activity. They could observe the goings-on—observation is a favorite recreation for all cats—they could interact with the people, and they could interact at least visually and vocally with the other tigers. The relatively smaller cages didn't bother them—they lay calmly on their shelves and didn't pace but spent their days participating, sometimes as the audience, sometimes as the players, in the very active scene of the tiger lab. In short, they had a life. Dr. Tilson pointed out that all through the study the tigers had a very healthy aura about them. Their mood was high, so they looked well. Their coats were smooth, their eyes were bright, their gums were clean, and their teeth were shiny. In fact, their overall health was better than the health of the tigers in exhibits. Best of all, from the point of view of the scientists, was that the blood chemistry profiles of these tigers gave positive proof of their well-being.

Perhaps it is possible for us to imagine the differences in the life-styles of captive tigers, or at least to imagine what life might be like for us under similar circumstances. Comparable to a tiger's life in the wild would be the normal life of a person who has a home and family and who goes to work every day to put food on the table. In contrast, prison is the only life lived by human beings that even remotely compares to life in a zoo—prison, or the room with the yellow wallpaper made famous by Charlotte Perkins Gilman. To better imagine zoo life, you might picture yourself living with your brother (if you are male) or sister (if you are female) in a department store's window display that looks like a luxuriously furnished home. Satin drapes shroud the French doors, white woolen upholstery encases the armchairs and the sofa, and a thick silk Oriental carpet covers the parquet floor. But the doors lead nowhere, the books on the shelves are fake, the TV doesn't work, the radio has no innards, and the only magazine, a copy of *House Beautiful* on the coffee table, is dated 1980. Anyway, you have read it so often you now know it by heart. Long ago you and your sibling have resolved all your differences. You have little to say to one another and you no longer think of escape. You have forgotten your freedom and have accepted your fate. The building is your prison, and both of you realize that you will never leave it alive. To forget the boredom and the crowds of people going freely wherever they please, who gather each day outside the glass window, oohing and aahing at the luxury that surrounds you, you and your sibling lie down behind the sofa, where you escape into dreams. You don't wake up if you can help it, not even when people in the crowd notice your feet poking out beyond the sofa and bang on the glass to rouse you. You dream of the night, which you spend with three or four other prisoners shackled to the chairs in the employees' lounge. At least you and your fellows can talk all night without wild-looking faces staring at you.

But imagine yourself in a circus. You are often cold, you are sometimes hungry, and your quarters are no bigger than the inside of an automobile. But you have plenty of pleasant, friendly communication with your keepers; you have plenty of interplay, friendly and otherwise, with all the other prisoners, and you always find a lot to watch—people and animals constantly coming and going. Most of all, you know that three or four times a day you will be called upon to do demanding work that requires both mental and physical skills. True, crowds of people watch you do this, but the crowds look like a dim wall far away, a wall that breathes and vocalizes every once in a while but doesn't bother you. And at the end of the day when the work is done, you know you'll sleep in peace, secure in your own small space where no one can get in to harm you.

And thus, incredible as it may seem, as far as tigers are concerned, the circus is not a bad way of life. Contrary to some erroneous statements put out by well-meaning but seriously misinformed animal-rights activists, many circus tigers like their lives. They have meaningful work and are not tortured or hurt in any way by their trainers. This, too, is contrary to what the animal-rights activists would have us believe. But in fact, punishment is an ineffective way to train almost any living creature, and the good, skilled trainers don't use it. Training is best achieved by encouragement, practice, and rewards. The more sensitive the trainer, the easier the job. Many animals are very fond of their trainers and look forward to sessions in the ring.

All this is not to say that no trainers are cruel. Some certainly are. And some caged tigers are kept in appalling conditions. I once learned of a man and his wife who got tired of their menagerie of tigers and simply abandoned them. Taking their household items, the couple got into their car and drove away for good, and the tigers slowly died of thirst and starvation in their cages. However, that such things can happen is certainly not a reason to prohibit animals from performing in circuses, a remedy that some animal-rights activists advise. Terrible things happen regularly to pets and farm animals, but no one suggests that we abolish pet ownership or farming. Better to spend time and effort preventing the abuse. Anyway, at present, it is much easier to catch and punish a cruel tiger owner than a cruel farmer or pet owner, at least in the United States, since all captive tigers must be registered and since the conditions of captivity for all large carnivores are governed by a number of federal and local laws. The laws are designed largely to protect the public, but not entirely, and the effect is often the same.

Yet it is the tigers themselves who attest to their good treatment. I have watched an uncounted number of circus acts and over a four-year period have spent much time observing the training sessions and performance sessions of nine trainers and their tigers and/or lions. I've watched little traveling menageries as well as such giants as the Ringling Brothers Barnum and Bailey Circus. Usually the tigers in question stand patiently by the doors of their cages waiting to go into the ring for a practice session. Sometimes their faces brighten in anticipation when their trainer enters the barn.

This is not to say that tigers are never frightened before entering the ring. Often they are, but the cause of their fear is almost always another tiger or an unfamiliar piece of equipment. On one occasion that I know of, it was a large collapsible pyramid, and on another, it was a huge mirror-spangled globe on a rotary turntable that groaned as it revolved. On their first encounters with these mechanical horrors, the tigers shrank in fear to the far side of the ring. It was then the trainer's task to accustom them to the machinery, which in both cases he did by luring them forward with little bites of meat. By the third or fourth session, the tigers were walking up to the machinery voluntarily, and by the fifth or sixth session, they were using it.

Fear of being out in the open is another cause of tiger

malaise, and indeed of malaise in all cats. Tigers try to overcome this by skirting the edge of the ring. Fear of other tigers
is probably not so easily assuaged. The tigers with which I was
most familiar could become quite frightened of each other, but
not consistently so, which showed, I thought, an ongoing but
unstable relationship. In each case the trainer took great care
to watch the threatening tiger carefully, so that the threatened
tiger didn't feel called upon to defend himself or herself. A feeling of give and take existed among these particular tigers, so
that no one tiger seemed to dominate the others, or not consistently, or for long. In contrast, these same tigers—even the
biggest and fiercest of them—seemed in mortal fear of a certain lion who belonged to another trainer but was sometimes
quartered with them. Never did these tigers and this lion get
into the ring together, yet the mere presence of the lion in the
tiger barn could get the tigers bouncing off the walls of their
cages. Why? No one knows why. The lion had never been caged
together with any of the tigers and had never harmed one of
them. But he looked frightening and he acted frightening.
Probably he wanted to seem frightening. And who could blame
him? As the male of a species whose best strategy is to live in
pairs, with either a brother, a littermate, or a friend, this lion
lived all alone in a roomful of tigers. They must have seemed
to him to be in a group, and he wasn't. If the cages were opened
and they all came out, they would number in the dozens, but he
would have been all alone. Perhaps that is why he always assumed an intimidating demeanor. He never walked into the
barn if he could charge in. He seldom hummed or moaned, but
he often roared. One day he roared every fifteen or twenty minutes, thirty or forty times each session. The tigers hated to hear
it and stayed absolutely still.

Also he sprayed. His cage was on the corner with two sides
exposed to the room, one side exposed to the runway, and one
side exposed to another cage which was empty because no tiger
wanted to live next to that lion. The first day I observed him he
had just arrived and was making sure that everyone knew he was
present. On two sides, from the floor almost to the ceiling, the

walls were dripping with urine. And as I was standing there scribbling notes and wondering why he felt compelled to spray so frequently, my glasses suddenly went blank, like a windshield when the car before you speeds through a puddle. And my clothes and hair were wet—he had sprayed me!

But I've never seen a tiger show fear of a trainer. Nor have I seen a trainer show fear of a tiger. On the contrary, most trainers enter the ring carrying only a pole and a wand. The pole has a point on the end and is used for delivering small bites of meat to the tiger's mouth, and the wand, to which a length of string is often tied, is whisked under the tiger's chin to get his attention or, because he draws back from it, to persuade him to lift himself up. The wand is certainly not used to beat or whip the tiger, since one of the important bits of information that a trainer must impart to his tigers is that a training session is for pleasure, for success, and for many little snacks of meat, not for confrontation or fighting. In the former atmosphere, it is the trainer who prevails. In the latter atmosphere, it is the tiger.

In short, training often pleases tigers because, along with the snacks, it gives them something to do, which explains why many circus tigers seem vastly more alert than many zoo tigers. Circus tigers are also more responsive, more expressive, and more vocal. And they seem to live longer than zoo tigers.

Perhaps this is only because they are kept longer. A good, cooperative, well-trained tiger with long experience and ever-increasing skills is even more valuable in old age than in youth, something that cannot always be said of a zoo tiger who has contributed his genes and is simply taking up cage space. But probably circus tigers live longer than zoo tigers because they are happier and more stimulated. Except on rest days, circus tigers rarely pace or enter the comatose state that seems to be the norm for zoo tigers. Circus tigers are entertained and have a reason to live. Most zoo tigers are bored, and they show it.

My own brief and very informal survey of tiger longevity revealed a small but perhaps significant difference between zoo and circus tigers. In zoos that did not participate in the endangered-species breeding program (of necessity a life-shortening affair), the average age of large zoo cats was nine. But in one circus act, the average age of the tigers was thirteen, and in another, eleven. Recently deceased zoo tigers whom I happened to hear of had lived only eight to ten years and then had died of unstated causes. In contrast, numerous circus tigers were still going strong at fifteen and sixteen, with some continuing in good health into their twenties. A circus lion, Ace, lived for twenty-seven years as the much-beloved partner and companion of the trainer Klaus Blaszak. The relationship between Ace and Klaus was such that Ace would sit next to Klaus at mealtimes and steal food from his plate. When I met Klaus's wife, the trainer Ada Smieya-Blaszak, and their son, Brunon Blaszak, also a trainer, they had in their collection a number of elderly tigers including a tigress named Rowena, who was seventeen, and a tiger named Harry, who was fifteen. Despite her age, Rowena was the most agreeable of creatures, but being no longer young she would get tired late at night. By the time the last show started at 11:00 P.M., this elderly tigress would be asleep in her cage. When the music would start for the tiger act, she'd wake up and come willingly into the ring, since she was a loyal and veteran performer, but she couldn't wake up as well as she once could, and she would sometimes doze on her stool. Her eyelids would close, her head would droop, her thighs would loosen, and her knees would slowly spread. At last, even her tail would lose its tension and would hang straight down. The trainer and the other tigers would then exchange a glance. All would realize that the act was about to continue without Rowena. Sometimes she'd start to topple and would wake with a jerk, and if her cue was coming she'd try to stay awake to do her part, but once back on her chair she'd doze again. When the act was over and the spotlight had moved to the next arena, Rowena would rouse herself and trot appreciatively off to her cage, there to sleep until morning and to wake refreshed and ready for a new performance.

Harry was exactly the opposite. Age had made him angry and disagreeable, so that he was much too dangerous to perform. Very few people were willing to get into a ring with Harry. Nevertheless, the Blaszaks did not kill him or sell him but instead took him wherever they traveled, giving him good care and an extra large cage because, unlike the other tigers, he didn't get his exercise in the ring. A large red, white, and blue tarp draped his cage to protect his privacy. Behind the tarp he would lie on his side with his raised head relaxed against the wall of his cage, blandly taking note of everything.

It is hard for most of us even to imagine a unit like the Blaszaks and their tigers, a unit that consists of four highly skilled people—Klaus and Ada and Brunon and his wife, Marita—and eight highly skilled tigers, each species cut off from contact with its own kind by the needs of the other species, all individuals closely linked, voyaging together through space and time like astronauts, like a wolfpack, like members of a combat platoon, or like a family.

To keep their tigers, the Blaszaks of course must accept the best of the jobs that come their way, which often forces them to live in some pretty bad surroundings. I met them at the Jolly Roger Amusement Park in the noisy city of Ocean Beach, Virginia—a triumph of strip development on which bars, hotels, motels, convenience stores, gas stations, and fast-food restaurants are jammed together by the thousand. To the east lies a beach, if you can find it underneath the tons of human flesh piled on the sand, and to the west lie the dirty waters of an oil-streaked harbor on which rafts of garbage gently bob. Perhaps the unusually large amount of garbage is due to an eerie absence of birds, particularly gulls and other scavengers—these, together with their nesting sites, have been systematically destroyed as part of a beautification program allegedly conceived by the chamber of commerce.

Not in this setting would one expect to observe tiger behavior. Yet in this very place I had the honor of watching a fascinating interplay between one of the tigers, Rajah, and his trainer, Brunon. Rajah was focused upon a nearby cage holding several pacing tigresses whose erotic moans bespoke their sexual receptivity. Rajah kept answering, needless to say, meanwhile trying different ways to reach the tigresses through the bars of his cage.

In the course of his daily routine, Brunon would pass by these cages, readying the equipment for the first show, strewing fresh shavings, feeding and watering the tigers. One would think that Brunon was showing the tigers only kindness, yet every time he passed, Rajah would leap at him, jaws wide, showing the length of his teeth and bellowing a horrible challenge. Why?

According to Ada, Rajah and Brunon had been youngsters together, and now that both were adults, Rajah saw Brunon as a rival for females. To put it very simply, Rajah was ascribing tigerish motives to a human being and was afraid that the young and powerful Brunon would take his women. Meanwhile, the behavior of the tigresses could only have further agitated Rajah. Whenever Brunon passed, the tigresses chuffed at him and lovingly rubbed their cheeks and chins on the bars of the cages.

Rajah repeated his threats all morning long, right up to the time he and the moaning tigresses were supposed to join Brunon in the ring for the first show. But amazingly, at that point Rajah's behavior changed completely, and his performance was adroit and smooth. Many people would not have been as professional as Rajah was that day—nothing in his bearing or his manner gave any hint of his differences with Brunon or of his feelings for the tigresses, for that matter. In fact, his good manners were such that when during the performance he felt the need to relieve himself, he stepped discreetly down from his stool and, while the performance swirled on around him, went to the edge of the ring, where he squatted, curved his tail upward, and defecated. Brunon and the other tigers ignored him, as polite people ignore a person who has excused himself to visit a bathroom. Finished, Rajah buried his scat with shavings by very slowly and deliberately scuffling, first with his right hind foot a few times, then with his left hind foot a few times, then with both feet alternating a few times. At last, in all dignity, his timing unblemished, he clambered back up onto his stool and waited for his next cue.

Scratching up the shavings may have said much about Rajah and his perceptions of his trainer, since he possibly saw himself as being on home turf in the presence of a more dominant male animal. Rajah might challenge Brunon vocally, but he evidently wasn't ready to stake a claim with a scat. If he had been, he would probably have left the scat uncovered. Equally interesting is that Rajah saw the ring as his home, even though over the course of time the tiger act would move from one place to another. Evidently the layout of the ring and cages and the dynamics of the act were what mattered to Rajah, not the larger setting. Other circus animals and circus people feel the same— witness the frequently told story of the ringmaster's dog who buries a bone under a circus wagon. The circus moves to a new town and sets up again in the same pattern as before, and the dog can't understand where the bone has gone when he looks for it under the wagon.

Traditionally, the cat act in a circus has had the appearance of a show of force, Man against Beast, whereby a man armed with whip and pistol subdued a beast armed with teeth and claws. To call these performances confrontational would be putting it mildly. But all that is changing. Today a more typical circus act is that of the trainer Eddie Schmitt, whose tigress, Natasha, has learned to ride the aforementioned rotating spangled contraption. Up she sits, her forepaws crossed and delicately lifted while, accompanied by lyrical music, she rides the slowly turning globe as thousands of tiny mirrors drench the audience with sparkles. The message is not that Natasha is dangerous but that she is beautiful.

Actually, however, she is both. This is what makes her so interesting. I think that many people, myself included, once found the confrontational style of cat act gripping, if only because the trainer seemed less like a tyrant coercing helpless animals than like a small person outnumbered by powerful creatures over whom he was keeping only minimal control. Yet even more gripping was the underlying message, that the entire spectacle was a performance, and that the trainer and his big cats were acting. There were just too many big cats in that ring for the snarling and whip cracking to be real. If the cats had wanted to maul the trainer, they would have mauled him, anyone could see, even if they themselves sustained a few casualties in the process.

So the modern circus message isn't really as new as it seems. Meanwhile, the old message undergoes its own metamorphosis. In recent years, every evening at sunset on a pier in Key West, Florida, a wonderful trainer named Dominique La Font performs with four graceful little housecats who open their own cages, leap fluidly onto their stools, and perform a series of breathtaking tricks, while Dominique, a chair in one hand and a whip in the other, pretends to keep them at bay as if they were lions. After the last performance, man and cats stroll off together, the cats to their carrying boxes, the man to his large RV in which they all live amid a pleasant domestic clutter of coffee cups, cat kibbles, and litter pans.

Zoos keep people and animals apart. Circuses unite them. In the Ringling Brothers Barnum and Bailey Red Show a man and a tiger mount an enormous double ferris wheel, the Wheel of Death, which lifts them higher than the aerialists, a process that evidently fazes this particular tiger not at all. Quite amazing. But for me, the most impressive moment of the circus comes just before, when the lights go down, and the tiger, leaving the center ring where the traditional tiger act has just been performed, goes to the ferris wheel. Loose, with nothing between him and the audience, he pads to his next act unrestrained and unguided except that his trainer walks at his side. This particular tiger wears a collar, and possibly is on a little leash, but the leash would mean less than nothing to a tiger of this size without the tiger's consent. He might as well be loose. Together the small trainer and his big colleague walk the length of the arena in the dark, the picture of professionalism. The lights go up when the two reach the ferris wheel. They both get on it, and the show goes on.

Rarely, the act itself illustrates the bond between an animal and a trainer. The most stirring circus act I ever saw was performed in the 1960s. The trainer was a beautiful woman with black hair, and the animal was a handsome lion with a black mane. These two also walked side by side and unrestrained from the door of the arena to a big swing, which they mounted. The lion sat on his haunches while the woman stood astride him, and then, in perfect silence, they began to swing. Higher and higher they swung until they were flying out over the heads of the audience. And at the limit of each arc the lion roared.

Do captive cats understand their circumstances? They want to. They try to. And with their awesome intelligence and their formidable powers of observation, they often succeed.

Perhaps the world's largest collection of captive tigers—up to seventy at a time including kittens—is kept in northern Illinois in a compound of barns and trailers enclosed by a heavy security fence. The facility is known as the Hawthorn Corporation, and in it, dozens if not hundreds of tigers are raised and trained to perform in circuses. The owner, John Cuneo, lives about twenty miles away, in the state's lake district, and was not present during my first few visits, but plenty of other people live within the compound walls—the grooms, the custodial staff, and the trainers, all under the direction of Roelof de Vries, who with his wife, Elke, have come from Holland especially for this purpose.

The tigers live in two huge barns in rows of roomy cages, one for each adult. Usually some of the interconnecting cage doors are open so that, except at mealtimes, two or more tigers can be together. The young adult tigers, especially those who are littermates, get along well and enjoy one another's company. Some of the older tigers, especially the older male tigers, prefer their solitude and are allowed to keep it. Many of the cages have access to outdoor pens where the tigers spend time in good weather. Each cage is also equipped with a high shelf where the occupant likes to rest, so that there is one shelf for each tiger.

Because these tigers have meaningful work which they seem to enjoy, so that most of them readily enter the ring for the daily rehearsals—eyes open, tails aloft and ears and whiskers forward—they are alert and lively creatures who assume the comatose state of zoo tigers only in the dead of night, when the barn is closed and all activity halted. Otherwise they are up and around, reacting to each other and to the ten or twelve grooms and trainers whose entire workday is spent in caring for them. The tigers, of course, know these people well and have pronounced personal relationships with each of them. The normal state of the tigers by day, therefore, is to be awake and alert, standing or walking through interconnecting cages, relating to neighboring tigers or to the people at their tasks, whom the tigers approach to greet, bowing their heads and chuffing as they do when greeting other tigers.

During my early visits to the Hawthorn Corporation I was of course a stranger, so the tigers seemed particularly interested in me, and whenever I entered the barn to visit their cages they would stop whatever they were doing to gawk at me. One day, therefore, I was mildly surprised to enter the larger of the two barns and see that the nearest tiger was high on his shelf, evidently watching something in the far corner so intently that he didn't notice me.

Accustomed to being greeted, I chuffed at him, but he merely glanced quickly to learn who had made the noise and then turned back, ignoring me completely. How strange. The tiger in the next cage didn't look at me at all, nor did the third. When I climbed a ladder that led to the top of the cages to see what was engrossing these tigers, I saw that all the tigers in the barn were on their shelves, all in the same position on their sides, all propped on their elbows, their heads high and their ears up. Quite obviously, all were alert to something in the far corner.

What was it? I couldn't see and didn't really like to climb to the walkway on the top of the cages, since whenever I had done so in the past, the tigers in the nearest cages would try to catch me—not that they could get their paws through the heavy wire net, but they almost could, and their efforts were unsettling.

The scene was most unusual. In the many hours I'd spent among these tigers, I'd never seen them do anything even remotely like this before. Yet I had seen enough of them to know what they were *not* seeing—I knew, for instance, that whatever they were looking at wasn't an animal such as a dog or a horse that would make them think of hunting. If so, some of them would surely have been crouching, tails twitching, and all would have been in different states of excitement. And I knew they weren't watching for their daily food, large slabs of meat which they had already eaten and which always arrived on a cart from the opposite direction. And they weren't witnessing the arrival of the lion or of another, strange tiger, since the arrival of a tiger would have had them on their feet, moaning and pacing, and the arrival of the lion would have had them bounding off the sides of their cages. Realizing that the nearest tigers, like all the other tigers, were so intent on watching whatever it was that they wouldn't come after me, I climbed higher and I saw—what? Just a group of four men standing together, talking. Three I knew—one was a groom and two were trainers—but the fourth was someone I had never seen before. Yet from the tigers' behavior I suddenly guessed who he was—John Cuneo, their owner.

That the tigers had been watching Mr. Cuneo was confirmed when he left the building. Although the other men stayed, as soon as Mr. Cuneo was gone the tigers turned from their vigil, got off their shelves, and resumed their normal activities. Immediately, the tigers nearest to me began trying to reach me through the bars. Later I was introduced to the man, who was indeed Mr. Cuneo. But I felt I already knew him.

I was deeply impressed by this episode. Considering all the people who go in and out of the barns and who deal every day with the tigers, why had they singled Mr. Cuneo from all the others for such rapt attention? Although Mr. Cuneo is an accomplished trainer, he doesn't train the tigers, and the tiger grooms, not Mr. Cuneo, feed and water them and clean their cages. So the tigers weren't getting their clues from any direct experience with him. Nor is he the only visitor who rarely appears yet is nevertheless very important to the facility. The veterinarians would also fit that description. No—how the tigers knew the importance of Mr. Cuneo has no easy explanation, but involves a process of fact finding and deduction on their part that is far beyond our present capacity to understand.

I n contrast to wild tigers, and indeed, to most wild members of the cat family, especially in the Old World, certain populations of cats in North America are doing surprisingly well. Partly this is due to re-population efforts, although not all of these have been successful. A scheme to reintroduce the lynx to the Adirondacks met with failure, presumably because Alaskan lynxes, who traditionally have enormous home ranges, were chosen to replace Adirondacks lynxes, who traditionally had quite modest ranges. Evidently the Adirondacks were not big enough for most of the Alaskan lynxes, who, when released, hiked far away. On the other hand, perhaps they merely wanted to return to Alaska. If so, they would not have been the first cats who tried to get home.

But perhaps the most dramatic recovery of the cat family in the New World has been the puma, who in parts of its former range is making an extraordinary comeback. Nowhere has this been more evident than in Colorado, in the Front Range of the Rockies, where pumas had been exterminated. Their demise had taken place at the turn of the century, the negative consequence of a gold rush with its swarms of prospectors and miners who, living off the land while working their claims, had in a very short time quite literally killed all the elk and all the deer and had eaten them. Of course the pumas, who were already besieged by Colorado's bounty hunters, couldn't live without their natural food, and they, too, vanished. Years later, more elk and more deer were imported and released so that the Colorado sportsmen would have something to shoot at, and the imported populations multiplied and eventually grew big enough to support a puma population too. Then the pumas returned.

The subsequent recovery of the puma population was enhanced by a change in the human demographics. The mountain men, who hunted, had been replaced by suburbanites, who jogged. And in keeping with the new consciousness along the Front Range, the state rescinded the bounty on pumas and instead levied a fine for killing them without a license or out of season. Under such favorable conditions, the puma population grew. Meanwhile, the suburbs of Denver and the outskirts of Boulder also grew, creeping along the canyons and up into the foothills of the Front Range into areas the pumas were also occupying. So today Front Range pumas appear in unexpected places and are even turning up in cities and towns.

Pumas are also returning in New England, to everyone's amazement. Officially, the eastern subspecies of puma, *Felis concolor cougar*, was eradicated over a century ago to make New England safe for farming. And according to many wildlife experts, the nearest surviving puma population is almost a thousand miles distant. Where, then, are the New England pumas coming from?

At first, the experts suggested that someone was releasing captive pumas. And because no responsible institution would do such a thing, the finger of suspicion pointed to pet owners. Yet to some, myself included, the release of pet pumas seemed more than unlikely, if only because no puma without claws could survive in the wild, and no puma with claws could be tolerated in a household. Like a housecat, a puma will sharpen her claws on furniture, but while a housecat would need to pluck for several years to complete the destruction of, say, a sofa, a puma could render a sofa to fluff in a matter of hours.

Even so, the New England authorities have generally denied the return of pumas and at the time of this writing continue to do so, albeit with difficulty. After all, New Englanders keep seeing them. So the only reasonable conclusion that can be drawn is that with or without the acknowledgment of the wildlife authorities, pumas have been here all along.

The early sightings of pumas in New England were discredited just as often as are the present sightings. During the early 1940s, many people in rural New Hampshire, myself among them, believed that a strange animal lived in the woods. Numerous people reported seeing a black panther, which eventually became known as *the* black panther. Since panthers were supposed to live in India or Africa, the individual in our woods was assumed to have escaped from a zoo. That the only zoos were far away in Nashua and Boston troubled nobody—being urban centers, Nashua and Boston were the source of everything exotic, and the idea of a panther from the city didn't seem at all strange. Still, not everybody believed there really was a panther. So sometimes someone brave enough to report a sighting was accused of poor judgment if not of drunkenness or downright misrepresentation. Probably a dog, probably a coyote, probably an otter, probably a housecat, the smug listeners would say.

With the death of the New England farm, and with the resulting armistice in New England's war on wildlife, reports of the panther grew more numerous and more far flung. The animal would have had to wear seven-league boots to get from one sighting to another. Anyway, by then the panther would have been over forty, or nearly twice the age of the oldest big cat known. At this point, it became clear that the creature people had been seeing hadn't been an escaped zoo animal after all but in fact had been several pumas.

The amazing thing is that there ever was any confusion. Pumas are huge animals. A full-grown male can weigh two hundred and fifty pounds, stand thirty inches at the shoulder, and measure nine or ten feet from nose to tail. And pumas scream. A dismal howl, a trilling wail, an unearthly scream, a low moaning cry, a loud wailing scream, a wild screaming cry, a loud weird cry, a loud penetrating scream, a queer half-human cry, like the scream of a terrified woman, like the scream of some woman in trouble, like the agonized voice of a boy or a woman, like someone screaming from the pain of surgery—these are some of the images that came to mountain men and other early settlers whose puma stories were compiled in 1946 by the early wildlife biologists S. P. Young and E. A. Goldman.* Why do pumas scream? They scream because they live far apart in the woods where trees prevent most sounds from traveling. If they don't scream they won't hear each other. So who would have guessed that any question would remain about the presence or absence of such a large, vocal animal?

Eventually I became so interested in the panther controversy that I went to New Hampshire's Department of Fish and Game in Durham, which, while denying the presence of pumas in the northeast, nevertheless maintains records of the sightings. There, in two thick files, I found accounts of more than a hundred sightings, beginning with the report of a puma killed in 1895 and ending with a sighting that had taken place just a few weeks before I read of it. Most of the accounts seemed reasonable and accurate, often mentioning features that are field marks of pumas, such as the account by one observer of a large cat with a long tail "as thick as a child's arm." That is an exact description of a puma's tail, and is especially telling because the only other cats that could be called large—lynxes and bobcats— have short tails. From 1980 to the time of this writing, no year has passed without at least one sighting. In 1985 there were seventeen reported sightings. Certain sightings seemed to cluster, as if several people were seeing the same puma, but even while allowing for this, the sightings increased slowly with time, probably an indication that a puma population was growing.

Even more fascinating were descriptions offered by ob-

*S. P. Young and E. A. Goldman, *The Puma: Mysterious American Cat* (New York: Dover, 1946).

servers who didn't understand exactly what they had seen. Often the fresh eyes of such an observer made the sighting particularly convincing, such as an account from 1984 in which the observer reported a puma being followed by a "little spotted dog." At first, the observation seemed flawed. Except for one or two rare and costly miniature breeds, there are no little spotted dogs. And such a dog would probably have more sense than to tag along behind a puma; if not, it would be killed. However, the description exactly fits a puma kitten, and the fact that the observer didn't know this made the observation all the more convincing. Furthermore, the presence of a kitten suggests two adult pumas and therefore some kind of community, not just a wandering stranger from afar.

Sightings continue, and the list of witnesses now includes a biologist with the U. S. Fish and Wildlife Service; also the former president of Radcliffe, Dr. Mary Bunting Smith; and finally—confirming a sighting in Connecticut—none other than Dr. George Schaller himself, perhaps the best and best-known wildlife biologist in the world, whose eminence began with his landmark studies of tigers and lions.

The testamentary evidence of such prominent people would have been quite enough for me, yet more was to come. On February 3, 1992, from my home in New Hampshire, I myself saw a puma in my own field. It was tawny, or deer colored, the color that was mentioned in most of the reports of sightings, and it was chasing something small that dodged. Around in a big circle ran the puma, then it vanished in the direction from which it had come.

Because I had just been writing about the controversy that swirls around pumas, and then had looked up from my desk to see one, I would have assumed I was merely hallucinating if Pearl, my dog, hadn't seen it too; she even seemed to know what it was, because she didn't rush out the dog door to launch herself at it as she would have done if it had been almost any other animal. Rather, she wisely stood rooted to the spot, but with her nose flattened against the glass window, her ears up, her eyes wide, and her body trembling with the strength of her emotions. An immediate search of the area by dogs and humans yielded nothing—the ground was frozen solid but was bare of snow.

More than excited, I telephoned everyone I thought would be interested, and we organized a real search party. Sue Morse, the forest ecologist and tracker, came down from Vermont to help us, bringing a pair of rubber puma feet to show the rest of us what to look for in a track. Ably assisted by many dogs, we combed large tracts of forest but found nothing, and the lures we set out—fish, musk, and catnip, promising the puma a chance to eat, have sex, and get high all in one place—did not, as far as we could tell, convince him. Through a friend of Sue's we even obtained a flask of puma urine (taking care that the donor was healthy and had been vaccinated so as not to make the local wild cats sick), but all we succeeded in doing with the urine, which we sprinkled in the woods, was to upset a certain bobcat, who, objecting violently to the perceived intruder, left his own emphatic scrape on top of the puma urine and marked the place with a scat. As a desperate, final measure, I bought a predator call, a device that makes an earsplitting shriek like a captive rabbit's, but, alas, even this did not produce the puma. Evidently, however, the puma was still around. A few days later a neighbor came by to say he'd seen a big, tawny, catlike creature in his field.

In contrast to the extreme shyness of the eastern pumas is the extreme boldness of the western pumas, who these days come right into the cities and towns. The newspapers of Denver and Boulder regularly carry accounts of pumas hunting people's dogs and cats, eating pet food left out on people's sun decks, and hiding in the dense, exotic shrubbery that people cultivate in their yards. In 1992 I made a study of the puma sightings that had occurred that year within the city limits of Denver and Boulder and came to the conclusion that the wanderers were young animals, probably off on their own for the first time and not coping well. The lack of a cultural taboo against our species, combined with inexperience and bad judgment, had brought them into the towns. Or so it seemed. In nearly every case, I had no difficulty figuring out where the youngster had come from or why he or she had chosen the place in question. All the places were at the edge of the foothills on the plains, and in all cases, I thought, the youngster, when starting his or her journey, probably would have had no choice but to travel east. West led back into the mountains, where, it seemed fair to conclude, the adult members of the growing puma population had acquired all the good territories. Surely big and powerful pumas already sat in the lookouts and patrolled the boundaries of the numerous Eastern Slope puma habitats and would chase any young newcomer away.

In fact, I could think of only one question that wasn't readily answered by topography combined with human demographics, and that was why the young pumas came so fearlessly straight into the towns when they could have slipped around them. In order to guess at an answer to that question, however, all I needed to do was to go up into the hills and see for myself what the youngsters must have seen as kittens. East of the foothills in the Denver-Boulder area, the Colorado plain is more or less a continuous urban sprawl, thick here, thin there, and from almost anywhere along the Eastern Slope one looks east at a sea of buildings with glittering windows, surrounded by streams of cars. The scene is sufficiently remote that sirens and horns and other urban sounds aren't audible; the scene contains, in short, nothing alarming when viewed by day. But viewed by night, it becomes positively entrancing for members of the cat family. Boulder, for instance, becomes a vast array of lights, some moving, others spinning, still others flashing, a scene so visually appealing to the cat family that it might have been designed just for their pleasure. As housecats watch television, not really caring what the program is about but thrilled by the irregular, jerking movements, so the young pumas in their mothers' lookouts in the Rockies might have viewed the flitting lights below.

The next step for the outward bound youngster is to start down the mountainside, toward the by now completely familiar landscape, which he or she has been viewing nearly every day and night since emerging from the den. The first human installations the young puma would encounter on the eastward journey would be the widely spaced homes of well-to-do suburbanites, who each year build houses ever higher on the slopes. Since, presumably, the appeal of such neighborhoods is their remoteness and their privacy, the human residents tend to be almost as quiet and as circumspect as the pumas themselves—after ten o'clock at night these neighborhoods are silent and dark, and most people are inside the houses. Often at night I've quietly walked through these neighborhoods just to learn what a puma might encounter, and the answer is: nothing. Needless to say, I wasn't furtive—on the contrary I walked in the middle of the road. But not one person ever saw me. No one even realized I was there. The dogs knew, of course—once in a while one of them barked, but usually from inside a house. In no case did anyone investigate, and sometimes the people even shushed the dog. So it seems that from the start a passing puma could expect no opposition.

Pumas like paths and follow them whenever possible. They have their own puma corridors, and they also follow our paths, trails, and roads, as well as riverbanks and canyons, much as we would. In consequence, most of the pumas who entered Boulder seemed to have come (or so I thought) along relatively few pathways. An informal and cursory survey yielded four possible routes into town that I thought I could identify: two of the routes came down canyons and two followed the banks of brooks. Each canyon route led into an upscale neighborhood with neon green lawns and a puzzling maze of fences, but the brooks became unkempt, forgotten ditches and thus remained wild-looking even deep inside the city, which was where many of the sightings took place.

In most cases the puma wasn't sighted until afternoon. Then, the people who discovered him usually got excited, in contrast to the young puma himself, who usually remained perfectly calm. That, too, seemed disconcerting to his discoverers. A puma who shows no fear of people seems sinister, dangerous, especially when the people are so much afraid of him. However, the reason for the puma's calmness could also be surmised by traveling his likely route. Being of a crepuscular species, the puma would most likely start out on a journey sometime in the evening, after resting all day, just as we might start a journey in the morning after resting all night. That timing would bring a traveling puma into the city in the middle of the night, when the streets were empty and all was quiet. If a puma had tried to travel through the city by day, people would have seen him, with the resulting fanfare of police and emergency vehicles, inevitably ending in disaster for the puma.

So the puma surely arrived before dawn and found himself deep inside the city by the time the city woke up. Then all of a sudden, doors began to open and people and their dogs began to come out. Soon traffic filled the streets and the city was once again alive and busy. Surely the puma was taken aback, and surely he felt he had no choice but to hide as best he could. In the sightings that I investigated, most of the pumas had chosen to hide in bushes or long grass surrounded by open space—the most dramatic example being a puma that hid all day on the grounds of the school where my grandson, David, was attending seventh grade. From afar, that particular school must have looked wonderful to the young puma—set against a hill too steep for building, the school is surrounded by fields where the grass is long and golden, like wild grass. Best of all, a tangle of bushes grows at the edge of these fields. Hidden in these bushes, the young puma could watch the doings of all the people while hoping that none came too near or threatened him. After a day of watching warily as children swarmed around him, playing soccer and generally making lots of noise, he must have lost most of his fear just through acclimatization, and when in the afternoon someone noticed him, perhaps because he got up for a drink or to move into the shade, he had undergone about ten hours of city life during which nothing bad had happened to him. By then the busy school ground must have seemed somewhat less scary. Such was not the experience of his astonished discoverer, however, who probably had never before given much thought to pumas. No wonder the man was frightened when an enormous tawny cat calmly appeared from nowhere on the school grounds.

I didn't learn what happened to that particular puma. Possibly he was killed by the wildlife officers or by the police, although the attitude of the Boulder citizens is so favorable to pumas and, in fact, to most wildlife that the policy followed by the Colorado Division of Wildlife is to tranquilize the pumas and transport them back to the hills. Even so, the Division justifiably fears lawsuits and undoubtedly has to kill many straying pumas simply because of the vast number of litigious people who would view any encounter as a bonanza and sue the state.

Fear of lawsuits sealed the fate of a young puma who, in August 1992, somehow got himself into the heart of Denver. He had followed a dwindling watercourse to the intersection of Hampton and Monaco, two swarming thoroughfares lined with shopping malls and roaring with traffic. A worse place for a puma can scarcely be imagined. Even so, early one morning he entered a neighborhood next to these streets and hid there until eight o'clock at night, when a dog saw him and chased him into a tree. He had probably been trying to resume his journey. When the owners of the tree saw why their dog was barking they called the police. With the police came firemen, reporters, an entire television crew with vans, cables, lights, and cameras, and, eventually, an officer from the Division of Wildlife. The excitement drew a large crowd of passersby, who, much taken with the sight of the young puma peering down at them, begged the authorities to spare his life. The wildlife officer wanted to tranquilize the puma and move him out of town, but when the dart hit the youngster, he panicked and jumped out of the tree in a dash for freedom. The police opened fire and killed him.

He must have seemed enormous, at least to some of the reporters present. An early edition of one of the papers gave his weight at two hundred and fifty pounds. In fact, he was hardly more than a kitten, and in his starved condition probably weighed between sixty and seventy pounds. A photo in another paper shows him lying dead, his gaunt little body curled at the feet of the district wildlife officer.

After this sad drama appeared on the local television news, some people chose not to report sightings lest other pumas suffer the same fate. Such was the decision of a young couple in Boulder late one night when their dog, Bailey, chased a young puma up onto the roof of a toolshed. I met these people while investigating another sighting in the same neighborhood, and they told me that when they saw the puma looking down at the dog, they assumed it was trying to hunt him. In this they were surely mistaken. If, in fact, the puma behaved as they described—they said it had crouched on the roof for fifteen or twenty minutes looking down—it had been treed by Bailey and was afraid, nothing else. Yet, even believing as they did, the young people were generous to the puma and did not call the police. Instead, they brought Bailey inside and left the puma alone. In the morning, it was gone, perhaps headed back to the hills, and no harm was done.

Are pumas dangerous? Some are and some aren't. The pumas who find their way into cities have so far hurt no one, although dogs, cats, and chickens have evidently fallen victim to some of them. Even so, pumas do attack people. Pumas look for all the world like exceptionally beautiful housecats, so it is hard to believe that harm could come of them. Yet the fact that people are almost exactly the same size as deer cannot be lost on pumas. In other words, people are the perfect size to be prey, just as voles are exactly the right size for housecats, and attacks do take place. One June day, two pumas attacked a Boulder woman, after which one of them may have moved on to attack a young man. The Boulder *Daily Camera* of June 4, 1990, reports the first event, telling of the young woman who, while jogging in the hills near the city, saw a puma crouch low and come toward her. She threw a stone, but the puma kept coming. At that point, the woman noticed a second puma creeping up on her from behind, so, with great presence of mind, she made her way up an embankment and climbed a tree at the top. On her way up the tree, she felt pain in the calf of her leg, and looking down she saw that the pumas were climbing after her, one behind the other. The nearer of the two, the one who had scratched her, was looking up at her with its paw on her branch.

At this point, according to the *Camera*, the woman expected the pumas to kill her. However, refusing to give up without a fight, she stomped on the head of the nearest puma. That puma dropped to the ground. The other puma then climbed higher, snarling at her as she tried to drive it back with a branch. (An uncooperative attitude on the part of the prey sometimes angers a feline attacker—Rajah the housecat hissed and spat at a mouse whom he had carried indoors to play with, but who then evaded him by running into the toe of an overturned boot.) Eventually, though, the second puma also dropped to the ground. For a while the two paced back and forth under the tree; then they went off to drink from a stream and finally disappeared for good. The courageous young woman climbed down from the tree and ran a quarter of a mile to a group of houses.

A few weeks later, about twenty-five miles away, a puma who might have been one of the young woman's attackers killed a young man. He, too, had been jogging at the time of the attack, which came from behind. Searchers found his partly eaten body a short distance from the road; he had been dragged there and then covered with leaves by the puma as if he had been a deer. Eventually the searchers noticed a puma crouched at a distance, watching them. Later, this puma was hunted down and killed. It was a young male, about two years old, weighing about one hundred pounds. His stomach contained fragments of the young man's body, proving that the hunters had found the right animal and that a man-eater did not remain at large.

How many people have been killed by pumas? A variety of figures exists. The actual number, whatever it is, is very much less than a thousand, which was the number of pumas killed during the decade of the 1950s by one Robert McCurdy, a guide and bounty hunter in the Southwest whose tedious bragging is preserved in a blood-soaked biography called *Life of the Greatest Guide*. In virtually all countries, but certainly in the New World, far more people are killed by ordinary, domestic dogs than are killed by any kind of cat, still more people are killed by insects, even more are killed by lightning, and very many more are killed in sports-related accidents, to say nothing of the people killed by other people, especially by those with guns or automobiles. In contrast, we are surely the primary agent of death for all members of the cat tribe. For many if not most cat species, our depredations must surpass accidents, disease, and even starvation by a considerable margin.

Ironically, the very policies of game management, which in parts of the West combine a lax attitude toward poaching with a prolonged hunting season that encourages the killing of male cats, actually may contribute more to man-eating than any other factor. Why? Because Fish and Game biologists generally fail to recognize the social systems of the cats. Rather, many Fish and Game personnel cling to the superstitions that the wild cats are asocial and that male cats kill kittens indiscriminately. The fallacy of this belief cannot be overstated—if it were true, there would be no wild cats. In fact, the presence of a big male puma or bobcat or lynx or jaguar almost certainly serves to stabilize an area, because if he's already there, he is probably the owner, he has probably already committed the necessary infanticide, and at least for the near future he will probably keep any male newcomers away. Under his protection, the kittens he fathers will grow up and will stay with their mothers long enough to learn the necessary skills for survival on their own. Thus, under the protection of a resident tom, the cat population stabilizes or rises.

Many kinds of animal regulate their own populations, as of course they have done since their species began. This is particularly true of the carnivores. Yet the opposite belief, that human beings must control the populations of wild carnivores lest they overrun the food supply, persists like religious dogma, is a favorite recitation of furriers, and is even taught in certain low-quality biology classes. Thus the hunting of the wild cats continues, often on public lands where the game belongs to the federal government and is supported by taxpayers all over the country, not just by the hunters from the state. The big toms are shot by hunters and poachers, the once-protected ranges of these toms are opened to intruders, the intruders kill or disperse the dead tom's children, and the population drops.

Yet the fact that some of the dispersing orphans venture into cities does not suggest, even remotely, that most dispersing pumas cause problems to people or their pets. Most cats of any species and all ages remain shy and circumspect, even at the cost of their own lives. This is particularly true of young pumas. In just one year I learned of three different pumas, two females and a male, all between the ages of six months and eighteen months, who were found starved to death in the presence of mule deer and jackrabbits, not very far from human settlements. If these three were found at random, how many more had died that year who were not found? All were alone, and investigation showed no ailments, no injuries, nothing to explain the deaths except that they had not been eating. One, a young female, had the two wing casings of a beetle in her stomach. All three seemed to have been orphans, and all had died lying curled on their sides in long grass or under sheltering bushes, where, evidently, weakness overcame them.

Attacks on human beings by pumas were studied by Dr. Paul Beier from the University of California, who presented his findings at a conference on pumas sponsored by the Colorado Division of Wildlife in Denver in 1991. His study examined fifty unprovoked attacks that had taken place in twelve western states and in two Canadian provinces over the past hundred years. The most astonishing attack happened on Vancouver Island, which in itself was interesting, since the pumas of Vancouver seem more inclined to attack people than pumas anywhere else, suggesting, once again, a cultural bent. Also significant is that the inclination refutes a generally held belief that pumas are less bold where they are hunted. On Vancouver, pumas are heavily hunted by sportsmen, yet the pumas of Vancouver are said to be very bold indeed, and Vancouver continues to be the scene of an ongoing series of puma attacks on human beings. What, if anything, hunting has to do with this is unclear.

In the Vancouver attack, which was reminiscent of my friend Lissa's puma, Ruby, jumping at the deer in the diorama, a puma jumped through the large glass window of an isolated cabin, knocking over the only lantern and seizing the cabin's owner, a telephone linesman who was preparing for bed and had undressed to his underwear. In the dark the brave man, with the puma biting him, fought his way into his kitchen where he got a knife and stabbed the puma until it let him go. Running outside, he slammed the door behind him, closing the wounded puma inside and himself outside on a bitter winter night in the snow. Although badly injured, the nearly naked man got into his boat and rowed six miles against a strong wind and a heavy sea to a neighbor's cabin, where, since no one was at home, he broke in. By then close to death from shock and hypothermia, he had to huddle under a blanket for several hours before he could dial the telephone. His rescuers found and shot the puma, who was still locked in the cabin. Like most other pumas who have attacked people, this puma was a young male.

Of the fifty unprovoked attacks discussed by Dr. Beier, two-thirds were made on children. In eleven cases the child was alone, but in sixteen cases other children were present, and in eight cases an adult was present, all of which reinforces the notion that the size and age of the intended victim, rather than the presence or absence of another person, were significant to the puma.

In spite of so many attacks, a surprising number of people have no concept that pumas can be dangerous. Probably this can be explained by the way pumas look. Pumas seem very pretty to human beings and have the type of facial features often found on greeting cards—big eyes, round ears, and a small chin, all of which are guaranteed to reassure human beings. In this pumas are the opposite of wolves, whose facial features—long nose, pointed ears—frighten some of us. Some people react to the appearance of pumas very sensibly indeed, saying to themselves, Uh-oh—good thing the housecats aren't that big or we'd all be in serious trouble. Other people see what they take for cuteness and ignore the implications of the size. At the puma conference in Denver, a bizarre snapshot was presented. It showed a woman holding a baby while a puma peers around her, so that all three are looking at the camera, held by the baby's father. Evidently he thought it was safe to pose his wife and infant next to a large, wild, adult male puma who had been lurking in the bushes. In the photo, the wife is smiling and the puma looks puzzled. The puma's unsettling lack of fear should have raised a red flag in the minds of the little family but did not. All three escaped with their lives, as if by a miracle, but their misjudgment was on a par with that of another couple who, wanting a photo of their little boy being kissed by a bear, smeared jam on his cheek and shoved him out of the car. Going for the jam, the bear killed the child by accident, which of course resulted in the bear's death—she was shot by the park authorities.

(Animals who make such errors aren't always killed—the life of at least one Yellowstone grizzly was spared even though she had killed a photographer. She was saved by his photos, which showed how she had grown increasingly uncomfortable with his nearness and how she had warned him again and again to back off. Clicking away, he did not heed her warnings, and on that occasion, the authorities put the blame where it belonged.)

Meanwhile, an account describing similar naivety appeared in the August 21, 1991, *Wall Street Journal*. A committee was formed on Fire Island, in Long Island Sound, with the intention of introducing "thirty breeding pairs" of pumas to control the deer that overpopulate the island. Credit must be given to the committee for attempting an organic solution to a problem involving wildlife, but the plan would not have worked. Even if Fire Island were good puma habitat, which it is not, it might be big enough for one puma, but not for sixty, as the committee proposed, and certainly not for two hundred pumas eighteen months later if the original pumas bred as planned. Possibly no one had read the puma literature, which reveals that cats don't form breeding pairs; and surely no one had bothered with the math, which suggests that by the fifth year the swarming pumas would have required no fewer than fifty thousand deer, and, without a practical way to leave the island, would be looking around for another source of food. The mauled and sorrowing citizens might then hold the committee responsible for the misfortunes that would result, but that possibility didn't seem to have been visualized either, perhaps because the committee was under the grave misimpression that pumas "present ABSOLUTELY NO DANGER to humans or their pets," as a poster proclaimed.

One result of the conference in Denver was that a sort of protocol was endorsed to encourage people to make their property less hospitable to pumas. Homeowners were advised to eliminate hiding places by cutting back their shrubbery, and to eliminate sources of food by bringing pets inside at night and not feeding the deer. Hikers were told of a canned pepper spray to be aimed at the face of an attacking puma but were warned against spraying into the wind, which could leave the hikers thrashing helplessly on the ground, unable to see or hear, right in front of the advancing puma. (When I tested a can, the product malfunctioned, letting fall a drop or two of peppery juice on my feet. What a disappointment that would have been if I had been trying to deter a puma!) More reliable methods for repelling puma attacks make use of the fact that pumas are sometimes susceptible to a display of force. If attacked, one might shout in a deep voice or make threatening gestures or brandish a stick. The key word here is *sometimes*, and the point is illustrated by the experience of a hiker who, against park regulations, had unleashed her miniature terrier at the head of a trail that led down a narrow canyon with steep walls. As the two entered the canyon—the dog first, the woman second—a puma, who evidently had been watching them from a ledge on the canyon wall, suddenly plopped down in front of them. The brave dog flew at the puma, who seemed aghast and fled along the trail with the terrier at his heels, demonstrating that even a tiny creature can route a puma by showing some convincing aggression.

On the other hand, one shouldn't count on it. Rethinking the situation even as he fled, the puma suddenly spun around, seized the terrier, and leaped up the canyon wall with the dog in his mouth. Neither was ever seen again.

Of course, the more people present, the better the chance of spotting the attacking cat in the first place. Like Ruby in her mock attacks on Lissa, like most man-eating leopards and tigers who attempt to prey upon human beings, man-hunting pumas tend to attack from behind, with a bite to the back of the neck. According to Dr. Beier's survey, most people who were attacked by a puma and lived to tell the story reported that they never knew what was happening until they felt the terrible thump of its body, and then the teeth and claws.

Even so, some people were able to fight off the puma. According to Dr. Beier's survey, of the thirty-five children who were attacked, nine were alone and were killed, but the others were saved because someone else saw what was happening and came to the rescue. Also according to the survey, once the puma had launched its attack, fighting back turned out to be an effective form of defense. Playing dead, which seems to help people when attacked by bears, was strongly counterindicated in a puma attack, as the deception could result in the victim's being dragged to a more private location with fatal consequences. After all, the bear's aggression often results from his fear of the human being, who allays that fear by assuming a nonthreatening posture. The cat, in contrast, may be looking for a meal, and would find a nonthreatening posture inviting. When attacked by a cat, it would be much better to do what bears themselves do in tiger country—stand up tall and face the attacker. One might stare at the cat's eyes and shout "Bad puma!" or "Back off!" in a deep, menacing voice while brandishing something—one's jacket, say, or one's camera. The cat is only trying to eat, after all, and doesn't want trouble. But don't stoop down to get a stick or a rock—children and crouching people are the most frequent victims of puma attacks. Why so? Because most hunting, or hunting in the old way, is pragmatic food-gathering behavior and is best done carefully, wisely.

Hunting for food (a different activity than hunting for sport) lacks the aggression that drives reckless, warlike behavior. One of the primary rules of hunting for food, at least for the animals and people who live in the old way, is simply this: be careful, tackle only what you think you can handle, and above all don't get hurt.

One of Dr. Beier's accounts of escape from pumas brought much laughter and even derision from the audience, as if he had intended the story as humor, which as far as I could tell he had not. I repeat the story here because of its implications and because, to me, it is the most intriguing of all the accounts. A woman who was backpacking alone on a remote trail was attacked by a puma. In the usual way, the puma leaped on the woman from behind, grabbing her by the pack and knocking her down. But she twisted herself around in her pack straps and kept the pack between herself and him. Face to face with her attacker, she spoke smoothly and encouragingly. Although he had her in his power and could have killed her at any time, he instead was willing to listen to her. Eventually, other hikers came along the trail and drove him away.

Why did this amuse some members of Dr. Beier's audience? Perhaps because the concept of making an emotional or intellectual connection with a puma at such a moment was remote from their experience, and they were unable to see themselves in a similar situation, to make the leap. "Nice kitty," some people joked, as if they thought that the puma had attacked because he was angry and that the woman had managed to placate him.

In fact, something very different was probably the case. Probably, the woman's voice and demeanor had taken the puma by surprise, so that his hunt metamorphosed into a different kind of encounter, requiring different behavior. He must have been wondering what that behavior should be, and in hopes of getting more information kept watching and listening.

After all, any intelligent, empathetic social being must often make decisions about his or her relationship to another being. Is the other being a friend or a foe or a meal? Because cats are creatures of the edge, dependent entirely upon animal protein, they usually select the third option, even to the point of occasional, opportunistic cannibalism. But they don't always select the third option. The puma had been sensitive enough to see that the woman didn't act like meat.

The men and women of Dr. Beier's audience, mocking suburbanites who typically interact with the natural world only by occasionally going hiking or hunting or camping, were very far removed from this kind of dilemma. But the puma was not. Nor, evidently, was the woman. And at one time, when our kind lived in the old way, as wild animals still live today, neither were we.

The episode, I believe, shows something common to all hunting species, of which we are one. Wherever people hunt, similar episodes appear in stories. A hunter spares the life of a bird or a roe deer or a fallow doe, only to find that she is an enchanted princess. Usually he marries her, thus acquiring a beautiful woman and a kingdom too; not bad for a lowly woodsman. The only disadvantage of his kindness is that he loses the food value of her body, but in fiction that consideration would be crass indeed and of course is never mentioned.

In the real world, though, they who hunt for a living really need the food, and if for any reason they spare the victim's life, they go hungry. So they must squelch the tendency to empathize.

Long ago, the /Gwi Bushmen told the true hunter's version of the story, but that version has a cautionary twist to it. In the /Gwi story, a man takes the role of the hunting puma and a female elephant takes the role of the backpacking woman. The man has married the elephant, which worries her parents very much. They suspect his intentions and those of his relatives too. The parents are right to worry—the husband's younger brother is plotting to kill their daughter. One day she has a premonition of disaster and tells her parents that they may never see her again. Her husband's people are on the move, and after taking leave of her parents she follows. The younger brother looks back and, noticing that she is trailing them, waits for her to catch up. Then he tricks her into letting him kill her. Finished, he builds a fire to cook one of her breasts, which he eats, sitting on her body. When the husband looks back and sees his brother sitting high above the bushes, he fears the worst, and when he returns to find his brother sitting on his wife's enormous corpse, he is furious. But the younger brother hands him some of the roasted breast, which presently he eats. "You fool," jeers the brother. "You were married to meat and you thought it was a wife."

If cats could speak, they, too, might tell the /Gwi's cautionary version of the tale. Meanwhile, though, when they're not hunting they think and feel like the rest of us. We, too, must suppress the wish to empathize when we are hunting. At other times, though, cats and people are much the same. I saw this during my first visit to Lissa Gilmour and her puma, Ruby, when chance provided me with an interesting tape recording. Ruby was sitting with us in Lissa's living room, where, in a confined space, the puma seemed much larger and more daunting than she had seemed outdoors. Worse yet, she seemed restless. While trying to keep track of her whereabouts and her changing moods, I found that I was missing much of the fascinating information that Lissa was sharing, so I got out my little recording device. Intended as a dictation aid, it records at an unnatural speed to save tape, making the words hard to understand and unpleasant to listen to. Tones, however, come through clearly, and, stripped of their purely logical sense, they sing of our thoughts and feelings, revealing far more than our words do.

At the beginning of the tape, Lissa and I are seated on her rug, with Ruby stretched out full length between us. Lissa and I are getting to know one another. I am excited to be with Lissa in her houseful of wonderful animals, and, as the tone of my voice attests, I am eager to please her. She, in turn, with her flawless western hospitality, seems happy to discuss her beloved animals with me. Our voices are soft and high, our speech is quick and often punctuated with brief stops—we are communicating enthusiastically but at the same time we are anxious not to interrupt each other. We do, though, and as women will, we keep laughing politely and saying, "I'm sorry!" Together we are singing a duet, the female version of the human species song.

Ruby, meanwhile, has found an odor on the rug. It is a drop of amniotic fluid from Lissa's pregnant Himalayan cat, Yehti, who, it turns out, has been in labor all this time. However, at the moment in question no one knows this, as Yehti is holding back her kittens, possibly because she is afraid of Ruby. Ruby does nothing to reassure her. Rather, she begins to rumble, then to spit. On and on she goes, sometimes so loudly that she drowns out other voices. She is singing the murderous song of a cat's dark feelings, a contralto solo of envy and displeasure, which fills Lissa's house and finds its audience of one, little Yehti, crouched in hiding under a chair, quietly considering the seriousness of her situation.

Suddenly, Lissa's canaries begin to sing. They are males and rivals, whose burst into song, within a nanosecond of each other, seems miraculously coordinated, as if both had been on edge and on the mark, waiting tensely for a signal. Surely they are the most accomplished of finches—their song soars high above our heads, filling the air with an unquenchable cascade of music. Once started, each bird with head thrown back pours forth song as if to extinguish the other.

The rest of us don't seem to hear the canaries. Lissa and I keep right on talking, so involved are we with the relationship of guest and hostess that is developing between us. Prowling Ruby keeps up her disconcerting spits and growls. Thus the tape brings out the voices of five creatures—two people, one cat (just

Ruby, because Yehti keeps quiet), and two canaries—each with its important, earnest message to the other of its kind.

As the afternoon passes, however, the canaries subside, and the rest of us—all but Yehti—seem to turn outward, we toward Ruby and she toward us. At first, I may have been somewhat frightened by Ruby. Well, maybe not exactly frightened— Lissa's confidence was reassuring—but Ruby was bigger than me, vastly stronger, a stranger, and a cat. For my part, I had never before been so close to a puma. Nor did I know very much about pumas. So I was cautious. On Side B of my tape, I hear my uncertain voice telling Lissa that I respect Ruby.

Meanwhile, I am also very curious about her. Pumas are considered to be small cats, but how close to housecats are they really? I hear myself asking if Ruby eats crouched above her food in small-cat fashion, or if she stretches out beside it, the way the big cats do. Here Ruby, who had been prowling the room, turned and walked toward us, eyes front, as if she had every intention of passing between us, when plunk—as if her rump, not her head, had made the decision—her hip hit the ground beside us, gracefully followed by the rest of her. She then twisted onto her back and showed us her furry white belly. On the tape I am asking how many breasts she has. Has she six, like a housecat? Or four, like a leopard or a lion? Actually pumas have eight, but I don't know that at the time, and as Ruby remains on her back I ask if I can search for her nipples. In attentive silence, Lissa and I then do just that, with Ruby lying flat on her back very stiffly, her head raised, her paws bent at the wrists, her thighs spread, looking down at herself uneasily while our hands roam like spiders through her fur. However, as Lissa quietly predicts on the tape, we find nothing. Maiden pumas have absolutely no sign of nipples or breasts. This in itself is fascinating. I keep searching, but when Ruby's body tenses, I stop fast. A housecat at that point might have seized my hand, clapped it to her mouth, and bitten it, and I don't want that treatment from Ruby, whose eyeteeth seem as long as my fingers and whose triangular cheek teeth are as massive as my folded thumb.

* * *

In the late afternoon, Ruby grew more restless. Soon she was jumping on and off the furniture, which she dwarfed. Next, she nervously paced to and from the window, growling and switching her tail. Judging from my voice on the tape, her agitation must have been making me nervous. I keep asking what Ruby is doing. Lissa's voice is very gentle and soothing and her words can't be deciphered. Soon a silence falls. Lissa has offered Ruby her arm, and Ruby, stretched at full length, has started to suck it. Her massive paws knead alternately, slowly, making bread. Ruby is like a kitten and has even aligned herself to Lissa by heading in to Lissa's side as a kitten aligns itself to the body of its mother.

In a low voice, Lissa asks me if I'd like to let Ruby suck my arm. I whisper yes, I would. Lissa shows me how to take her place in front of Ruby, and I do. My arm slips under Ruby's mouth. Feeling the change, Ruby slowly opens her great, yellow eyes and looks up, but by now I am deeply moved by her tenderness and vulnerability, and I just wait, not speaking. In peaceful silence, we all wait. There is no tension in the moment. Ruby again begins to suck. Then she purrs. The room fills with her purring. Little Yehti creeps out from under the chair and leaves the room. She will later deliver two kittens on a rug behind the bathroom door. Ruby doesn't care and lets her velvet eyelids shut. On the skin of my arm I feel the slow, gentle rasping of her rough tongue, which gradually turns as smooth and slick as a piece of raw liver. Has she turned her tongue over so that I am feeling the underside? Has she collapsed her papillae? I quietly mention this to Lissa, who nods. She knows what has happened, but not why. It doesn't matter. Ruby drowses, her black lips and smooth tongue gently pressed against my arm. And that is all. We stay a long time in the peace of Lissa's quiet room, not talking or whispering, just relaxing, purring, dreaming, gently breathing, mildly aware of cloud shadows, of the afternoon sun, of a light breeze from an open door, of being alive there together.

* * *

The next tape records the events of the evening, when we are in Ruby's pen, and Ruby is about to eat a pullet who has been accidentally suffocated by some other chickens. Crouching above the carcass in small-cat style, with her paws neatly together below her chest like a housecat eating from a dish, Ruby is preparing her food. Little bones snap as she severs the wingtips, then feathers rip as she plucks the breast with short, neat tosses of her head. Feathers float around us on the evening air. Soon Lissa shows me something strange. Ruby is not spitting out all the feathers. In the sides of her mouth, damp feathers are clinging, positioned to go down her throat if she swallows. Why is this? If she eats the feathers anyway, why does she pluck them first? Doesn't she know about them? Can't she get rid of them? In search of further information I lie down beside the chicken to look up into Ruby's mouth.

Thinking back, I occasionally wonder at myself for that. I wouldn't have put my face so near the eyeteeth of a feeding dog, let alone a dog I had met so recently. But, unlike dogs, cats don't snap, and anyway, by then I trusted Ruby. And Ruby trusted me. No longer did we view each other as unpredictable strangers, or care how closely we approached each other's faces and mouths. Not for a moment did I think she'd bite me. Nor was she afraid that I might snatch her chicken. Thanks to the soothing, the bliss, that we had experienced earlier, we seemed to understand each other. We had crossed our species' boundaries and had found the common center in each other, where all creatures rest.

Acknowledgments

MANY PEOPLE HAVE HELPED ME TO GATHER THE MATERIAL FOR this book. I thank them all. Going cat by cat and beginning with the pumas, I'd like to thank Lissa Gilmour, the wildlife rehabilitator, and her puma, Ruby. My debt to these two friends is deep and very obvious. I wouldn't have attempted to write this book without first visiting them. I'd also like to thank Sue Morse, the forest ecologist, for her insights into tracks and tracking. For an unparalleled insight into puma behavior and habitat, and for the opportunity to see wild pumas, I'd like to thank Ken Jafek of War Eagle Outfitters in Malta, Idaho, and his colleague, Kevin Allred, as well as their extraordinary dogs. As Kevin once wrote for *Coonhound Bloodlines:* "Without [these dogs] there would be no [puma] study. They are always ready to go, even if it's two o'clock in the morning and 15 degrees below zero. Their devotion, courage, and desire is something to be respected." To that I have but one word to add: Amen.

For information on puma populations and puma attacks, I'd like to thank James C. Halfpenny. For supervising a fascinating puma study, I'd like to thank John Laundre. For invaluable help in understanding the pumas of the Denver-Boulder area, I'd like to thank Kathy Green, Betsy Spettigue, and Mike Sanders. For further information on puma populations in the Rockies, and also for information on deer populations without which there would be no pumas, I would like to thank Allen E. Anderson. Let me hasten to add, however, that any mistakes herein on the subject of animal populations are certainly my own.

For generous hospitality, and for much insight into circus tigers, I'd like to thank John Cuneo, Elke and Roelof de Vries, Trudy and Bill Strong, and the tiger grooms and tigers of the Hawthorn Corporation. In this context I'd also like to warmly thank Harry and Rowena Thomas and their tigers; and to offer special thanks to Ada Smieya-Blaszak and her husband and son, Klaus and Brunon Blaszak, and their tigers, especially Rajah and Rowena; and finally, I'd like to express my gratitude to Rodney Huey of the Ringling Brothers Barnum and Bailey Circus for his help and his courtesy. For much insight into zoo tigers and for generously explaining the captive breeding program I'd like to thank Dr. Ronald L. Tilson, director of biological programs at the Minnesota zoo.

For information about domestic cats I'd like to thank certain cats, namely the hefty, part-Siamese barn cat Manas; the wise, marmalade tomcat Rollo; the elegant tricolor Baby Cat, and her marmalade colleague Thomas; also the blind white cat Aasa (now Chushi); the white hunter Orion; the black huntress Wicca; the elegant Rose and her sister, Iris; the stealthy Goniff; the cheerful Max; the devoted Natasha; the loyal Fang, a neosaber-tooth, and also his associates and rivals, the cats who roam the short-grass plains of Marshall Lane in Austin, Texas; the vanished Fritz, whose owners are still searching; the shy Cleo; the valiant Eddie; the gray Christmas with her pure white stockings; the mauve, mottled Lilac; and the black wondercat Rajah, all for the insights they have afforded to anyone who would take the trouble to watch them over time.

For information about training housecats as circus performers I'd like to thank Dominique La Font and his very accomplished cats, Piggie, Sharkey, Spot, and Mars, whose wonderful work can be seen in Key West on Mallory Pier every evening at sunset; for his observations of housecats hunting deer I'd like to thank Dave Blanchette; for the story of a cat protecting a person, I'd like to thank Lisa Rappaport; for the story of Bubastis and the tomcat who climbed high buildings to find her, I'd like to thank Mab Gray of the publicity department at Houghton Mifflin; for the story of Wazo, I'd like to thank Margie Bourne; and for the story of the cat who soothed a person, I'd like to thank Lori-Ann Tessier.

For insight into Ju/wa hunting and for observations about the Bushmanland lions I'd like to thank my brother, John Marshall. As has been seen in the text, many of the most interesting observations included herein were made by him. I'd also like to thank Tsamko Toma and his late brother, /Gashe Martin, as well as their late father, /Toma—all three of /Gautscha in Bushmanland, Namibia—for many important insights and much information about lions over the years. I'd also like to thank the wildlife biologist Richard D. Estes, who constantly and generously shared his vast store of knowledge on the behavior of African and other mammals. For insight and advice about the Ju/wasi in general and about their relationship to lions in particular, I'd like to thank the *grande dame* of hunter-gatherer anthropology, Lorna Marshall. For insight into circus lions, I'd like to thank Trudy Strong and also Timba, a lion.

For including me in their research project on elephant vocalizations during two seasons in Namibia, I'd like to thank my dear friends Katy Payne and Bill Langbauer. If not for the time in the field spent with them I would never have been able to observe the cultural change exhibited by the local lions. Furthermore, Katy is as attuned to animals as it is possible to be—in this she's more like some of their kind than like some of our kind. To spend time with Katy in the presence of animals is an unforgettable experience, of which I've had many, and I also thank her for those.

In a slightly different form, much of the information herein about lions first appeared in *The New Yorker*. For this I'd like to thank Bob Gottlieb, and also Nancy Frankin and Hal Espen, whose skilled and careful work so greatly enhanced mine. For his help and friendship, as always, I'd like to thank Ike Williams, my agent. For the beautiful line drawings that so sensitively link this present work on cats to my past work on dogs, I'd like to thank the artist, Jared Williams. For her friendship and for being so pleasant to work with as well as for her wonderful editing skills, I'd like to thank my editor, Becky Saletan. For her generous help and her ability, I'd like to thank my assistant, Anita Marie Mann. Finally, I'd like to thank Irene Williams and Mab Gray of the publicity department at Houghton Mifflin for the wonderful work they did in publicizing an earlier book of mine called *The Hidden Life of Dogs*. Their excellent skills on behalf of that book have paved the way for this book, even though the two books were published by different houses.

Bibliography

The following is a partial bibliography of the publications I found most helpful:

Clutton-Brock, Juliet. *Cats: Ancient and Modern.* Cambridge: Harvard University Press, 1993.

Cole, D. D., and J. N. Shafer. "A study of social dominance in cats." *Behaviour* 27 (1966): 39–52.

Corbett, Jim. *More Man-Eaters of Kumaon.* London: Oxford University Press, 1954.

———. *The Man-Eating Leopard of Rudraprayag.* Suffolk: Richard Clay and Company, 1947.

———. *Man-Eaters of Kumaon.* New York: Oxford University Press, 1946.

Dunstone, N., and M. L. Gorman, eds. *Mammals as Predators.* Oxford: Oxford Science Publications, 1993.

Estes, Richard D. *The Safari Companion: A Guide to Watching African Mammals.* Post Mills, Vermont: Chelsea Green, 1993.

———. *The Behavior Guide to African Mammals.* Berkeley: University of California Press, 1991.

Guggisberg, C. A. W. *Wild Cats of the World.* New York: Taplinger, 1975.

Hornocker, Maurice G. "Stalking the Mountain Lion—to Save Him." *National Geographic* 136, no. 5 (1969).

Kitchener, Andrew. *The Natural History of the Wild Cats*. Ithaca: Comstock, 1991.

Leyhausen, P. *Cat Behavior, the Predatory and Social Behavior of Domestic and Wild Cats*. New York: Garland STPM Press, 1979.

Macdonald, David. *The Velvet Claw*. London: BBC Books, 1992.

———, et al. "Social Dynamics, Nursing Coalitions and Infanticide Among Farm Cats, *Felis catus*." In *Advances in Ethology*. Berlin: Paul Parey, 1987.

———, ed. *The Encyclopedia of Mammals*. Oxford: Equinox, 1984.

Marshall, Lorna. The *!Kung of Nyae Nyae*. Cambridge: Harvard University Press, 1976.

Mongtomery, Sy. *Spell of the Tiger*. Boston: Houghton Mifflin, in preparation.

Packer, C., and A. E. Pusey. "Cooperation and competition within coalitions of male lions (*Panthera leo*)." *Animal Behavior* 31 (1982): 334–40.

Schaller, G. B. *The Serengetti Lion: A Study of Predator-Prey Relations*. Chicago: University of Chicago Press, 1972.

———. *The Deer and the Tiger*. Chicago: University of Chicago Press, 1967.

Seidensticker, J., et al. "Mountain lion social organization in the Idaho Primitive Area." *Wildlife Monographs* 35 (1973): 1–60.

Sunquist, M. E. "The social organization of tigers (*Panthera tigris*) in Royal Chitawan National Park, Nepal." *Smithsonian Contributions to Zoology* 12 (1981): 239–41.

Thomas, Elizabeth Marshall. "The Old Way." *The New Yorker*, Oct. 15, 1990.

———. *The Harmless People*. New York: Knopf, 1989.

———. *Warrior Herdsmen*. New York: Knopf, 1966.

Tilson, Ronald L., and Ulysses S. Seal, eds. *Tigers of the World*. Park Ridge, N.J.: Noyes Publications, 1987.

Young, S. P., and E. A. Goldman. *The Puma: Mysterious American Cat*. New York: Dover, 1946.